Terrell Carver

ENGELS

Oxford Toronto Melbourne

OXFORD UNIVERSITY PRESS

1981

Oxford University Press, Walton Street, Oxford OX2 6DP

London Glasgow New York Toronto
Delhi Bombay Calcutta Madras Karachi
Kuala Lumpur Singapore Hong Kong Tokyo
Nairobi Dar es Salaam Cape Town
Melbourne Wellington

and associate companies in
Beirut Berlin Ibadan Mexico City

First published 1981 as an Oxford University Press paperback,
and simultaneously in a hardback edition

British Library Cataloguing in Publication Data

Carver, Terrell
Engels. - (Past masters).
1. Engels, Friedrich
1. Series
335.4′092′4   HX273.E56
ISBN 0-19-287549-3
ISBN 0-19-287548-5 Pbk

Printed in Great Britain by
Cox & Wyman Ltd, Reading

Preface

While there are many books on Marx and Marxism there are few books on Engels, and even fewer that take him seriously as a thinker. In this one I have attempted a close study of Engels's ideas. To a large extent I have allowed him to speak for himself, since his own words are suitably vivid. It has been my object to interest the reader in Engels's thought and its implications for contemporary social science and politics.

I am grateful to the University of Liverpool for granting me study leave to begin this book, and I am indebted to those who supported my application, as well as to my students there. I should like to thank Catherine Payne and Mary Woods for their careful and critical attention to my typescript, and Larry Wilde, Henry Hardy, Keith Thomas and an anonymous adviser for their very helpful suggestions.

I should like to dedicate this book to D.M.

Bristol   TERRELL CARVER
September 1980

Contents

Abbreviations

I have used three collections of the works of Karl Marx and Friedrich/Frederick Engels, since at the time of writing the *Collected Works* have advanced only as far as 1854. For these sets I have adopted a reference of the form: volume number, full stop, page number. Arabic, large roman and small roman numerals are used for volume numbers according to the following scheme:

Collected Works (Lawrence & Wishart, London, 1975–) have arabic numerals, e.g. 12.432 for volume 12, page 432.

Selected Works in two volumes (Lawrence & Wishart/Foreign Languages Publishing House, London/Moscow, 5th impression, 1962) have the large roman numerals I and II, e.g. II.432 for volume II, page 432. I have used this set because it contains material not included in the one-volume version currently in print.

Werke (Dietz, Berlin, 1956–) have small roman numerals for volume numbers, e.g. xvi.432 for volume 16, page 432. Where English translations were not readily available or were nonexistent, I have translated passages myself from this set.

Other abbreviations are as follows:

AD Frederick Engels, *Anti-Dühring*, Lawrence & Wishart, London, 1969.

C I Karl Marx, *Capital*, vol. 1, ed. Frederick Engels, trans. Samuel Moore and Edward Aveling, Lawrence & Wishart/Progress, London/Moscow, 1954, repr. 1974.

DN Frederick Engels, *Dialectics of Nature*, trans. Clemens Dutt, Foreign Languages Publishing House, Moscow, 1954.

SC Karl Marx and Frederick Engels, *Selected Correspondence*, trans. I. Lasker, 2nd ed., Progress, Moscow, 1965.

I have occasionally made slight alterations in the English translations listed above in the interests of clarity or accuracy. My insertions into quoted material are enclosed in square brackets.

The quotations from *Collected Works* and *Selected Works* are published by permission of Lawrence & Wishart Ltd.

1 Engels and Marx

Engels was a partner in the most famous intellectual collaboration of all time. Though on his own admission he was the junior partner, he was in fact more influential than his senior through his popularisations of the ideas of Karl Marx.

But Engels also had ideas of his own, and in this book I shall attempt to identify and assess them. Marx himself acknowledged a considerable debt to some of Engels's own works, and there are, of course, the famous works written by Engels jointly with Marx. I shall be discussing Engels's contribution to them, in so far as it can be determined.

For most of his life Engels pursued his own work and published it under his own name, and here we find the most difficult and important problems in connection with his thought. To what extent was Engels furthering the work of Marx in areas delegated to him by the master? Can Engels's independent works be read as if they were jointly written with Marx? Did Marx and Engels always speak with one voice, even when they wrote and published independently? The answers to these questions are important, because of the enormous influence exerted by Engels in person and in his writings on the development of Marxism, particularly in works which were widely circulated after Marx's death. In many cases these works were designed or assumed to be popularisations of works by Marx or joint works by Marx and Engels. Many socialists took these later works of Engels to be authoritative and definitive, and many conversions to Marxism were made almost entirely on that basis.

It is not a trivial question whether Marx and Engels agreed or disagreed on any particular issue, or whether their works contradicted each other or exhibited any significant differences. If there were any significant differences between the two (as I believe there were), then Marxism becomes a very complex phenomenon to describe, and all attempts to show it to be a monolithic, systematic world-view must be failures from the start.

Engels has not been badly served by biographers, who have given us two substantial works and a number of shorter summaries that tell the story of his life. What is lacking in the literature on Engels is a treatment of his intellectual life that is not always haunted by the spectre of Marx.

2 Journalist

Engels's early career was brilliant. At seventeen he was a published poet, and at eighteen a journalist so trenchant as to sell out a complete edition of a Hamburg journal. His 'Letters from Wuppertal', published in the spring of 1839, were a sensational attack on hypocrisy in the valley towns of Elberfeld and Barmen, the Rhineland district in which Friedrich Engels was born on 29 November 1820. Since the Engels family had been for generations well-to-do mill owners, the youthful Engels used a pseudonym. The identity of the correspondent 'Oswald' was not a secret kept from his friends, however, and once the confidential 'Ha, ha, ha!' was past, the deadly serious character of Engels's work emerged: 'everything I wrote was based on proven data which I have from eye-witnesses and ear-witnesses' (2.426, 446).

Engels used his own eyes and ears to good effect, and his portrayal of the physical and social circumstances of a small but intensely industrialised community was very sharp indeed. Pollution of the river Wupper by dye-works and of the inhabitants by drink set a scene of visual and cultural shabbiness: a Catholic church 'built very badly by a very inexperienced architect from a very good plan'; the columns 'Egyptian at the bottom, Doric in the middle, and Ionic at the top' that flank the ex-museum, now a casino. 'There is no trace here of the wholesome, vigorous life of the people that exists almost everywhere in Germany', Engels wrote, and the reason was factory work (2.7–8, 9).

Child labour, cramped rooms, overwork, consumption, terrible poverty, drunkenness, syphilis, lung disease, coal fumes, industrial dust and a lack of oxygen took their toll in the valley. Workers, Engels claimed, were divided into the decent and the dissolute; wealthy manufacturers, he remarked, 'have a flexible conscience'. Among factory owners, strict Christians 'treat their workers worst of all', cutting wages to prevent drunkenness yet

offering bribes at the election of preachers. Hypocritical Pro-
testants were the object of Engels's wrath: they exhibited a
'most savage intolerance . . . little short of the Papist spirit'.
Woe to a preacher 'seen in a frock-coat with a bluish tinge or
wearing a rationalist waistcoat!' Engels judged local preachers
to be ignorant folk and deplored their activities, which pervaded
and corrupted every aspect of life, not least the educational
system which he had only very recently left. One of the teachers,
so he said, was asked by a fourth-form pupil who Goethe was
and replied: 'an atheist'. Local journalists and poets received
their due, among them one Wülfing, 'a man of unmistakable
genius . . . his head crowned by a green cap, in his mouth a
flower, in his hand a button which he has just twisted off his
frock-coat – this is the Horace of Barmen'. The whole region,
Engels concluded, was submerged in philistinism (2.9, 10, 12,
17, 19, 23, 25).

Replying in an open letter to a critic of his articles, Engels
noted that he had 'throughout acknowledged competence in
individual cases', but that 'in general I was unable to find any
purely bright sides' (2.28). As an attack on provincial hypocrisy,
obscurantism, pretentiousness and bad taste the 'Letters from
Wuppertal' were extraordinarily vivid. An eyewitness account
of early industrialisation was firmly at the basis of Engels's
view, and this turned the work into something even more
interesting, and prescient.

The beliefs and interests of the young Engels were formed by
his family, schools and community in that as an adolescent
he reacted to them with intense hostility. His forebears had
been leading industrialists and worthies of Barmen and district since the days of his great-grandfather, a yarn merchant
and – characteristically for the area – the founder of works
for bleaching cloth and for ribbon- and lace-making. In
the second half of the eighteenth century the Wupper valley
became one of the most intensely industrialised areas in Germany. The oppressive philistinism of Engels's school and community were reinforced by pietism, a puritanical protestantism
revived after the French Revolution. Fundamentalist Christi-

anity could not withstand the discreet rationalism of some of Engels's grammar-school masters, however, and by the time he left school (just before his seventeenth birthday) his critical views were forming. During the year he spent working in his father's business Engels evidently perused such rationalist works as David Friedrich Strauss's *Life of Jesus*, published in 1835, which subjected the Gospels to thoroughgoing historical scrutiny. Then, when he was just eighteen, Engels left for Bremen to gain experience in the export trade. While working in the Free City he also drank, smoked, sang, fenced, swam, attended theatre and opera, got into debt, studied and did other things that young men do when they leave the provinces. He also made friends with liberals and radicals in the 'Young Germany' movement, which demanded an end to stuffy, self-serving conservatism in religion, literary criticism and politics.

Over the years 1839–42 Engels established himself as a political and literary critic in nearly fifty articles and pamphlets, among them an account of travelling steerage on a ship bound for America: 'a row of berths . . . where men, women and children are packed next to one another like paving stones in the street'. Here were the people, Engels remarked, 'to whom nobody raises a hat'; they made a sad spectacle. What must it be like 'when a prolonged storm throws everything into confusion'! (2.116, 117).

However, Engels had interests other than socially conscious journalism. While in Berlin during 1841–2 for his national service in an artillery brigade, he attended the university there as a non-matriculated student. 'Friedrich Oswald', social and literary critic, now took on theology and philosophy as new targets, defending the liberal, critical 'Young Hegelians' from an officially encouraged attack launched by Friedrich von Schelling, the professor of philosophy recently translated from Munich.

Ask anybody in Berlin today on what field the battle for dominion over German public opinion in politics and religion, that is, over Germany itself, is being fought, and if he has any idea of the power of the mind over the world he will reply that this battlefield is the University, in particular Lecture-hall No 6. (2.181)

G. W. F. Hegel's reflections on consciousness and being, history and the state, religion and nature, and a host of other topics too numerous to list were of monumental abstraction. Moreover some of his writings were ambiguous in that the conclusions he drew were not perhaps the only ones – or even the most defensible ones – that could be drawn from his philosophical analyses. His philosophy of religion, which lent itself to pantheistic interpretations, contrasted with his favour-able remarks on Lutheranism and public profession of it. Simi-larly, his justification of the Prussian state did not follow un-ambiguously from his reasoned consideration of economics and politics. Hegelians of the 1830s held varying views on these issues, but unsurprisingly they tended to hold them in certain combinations: orthodox Lutherans supported Hegel's favour-able comments on the Prussian monarchy; free-thinking critics of religion in general and Christianity in particular tended to be political liberals calling for representative government in Ger-many, though until a liberalisation of the press censorship in 1840 they had to do so discreetly. The latter views were held by the Young Hegelians, who flourished in Berlin and at other universities in Germany in the early 1840s. Engels seems to have read Hegel for the first time while he was living in Bremen.

Though Hegel had been dead ten years, he was in Engels's words 'more alive than ever in his pupils'; Schelling, by con-trast, had been 'intellectually dead for three decades'. The 'good, naïve Hegel' had believed in 'the right of reason to enter into existence', and the Young Hegelian radicals took this as their battle-cry. Schelling's view, according to Engels, was that Hegel's philosophy was 'just bits of nonsense which existed only in Schelling's head and laid no claims whatever to any influence on the external world'. Engels/Oswald and his Young Hegelian comrades opposed this view and were supremely con-fident: 'Youth has never before flocked to our colours in such numbers', and talent has never been 'so much on our side as now' (2.181, 186, 187).

An anonymous pamphlet, *Schelling and Revelation: Critique of the Latest Attempt of Reaction Against the Free Philosophy*, swiftly followed. Given greater space Engels set out the plain man's guide to the Young Hegelian movement in Germany. It is still a readable and reliable account, and much the most exciting. The principles of Hegel's own philosophy were 'throughout independent and free-minded', Engels wrote, but the conclusions were 'sometimes cautious, even illiberal'. The teachings of the great philosopher were 'conditioned partly by his time, partly by his personality'. His political views, and his philosophies of religion and law, suffered from an internal contradiction: radical principles and mistaken, conservative conclusions about society, Christianity and politics. The journals and works of the new, critical philosophers – Arnold Ruge, David Friedrich Strauss, Ludwig Feuerbach and Bruno Bauer – were enumerated and praised by Engels. 'All the basic principles of Christianity, and even of what has hitherto been called religion itself, have fallen before the inexorable criticism of reason.' Yet Schelling was called forth by the 'Christian-monarchic state' to defend orthodoxy in religion and politics. Engels thought that this defence was worthless: 'the first attempt to smuggle belief in dogma, sentimental mysticism, gnostic fantasy into the free science of thinking'. After a lengthy critique Engels advised his readers to 'turn away from this waste of time'. Hegel had 'opened up a new era of consciousness', and Feuerbach's *Essence of Christianity* – just published – was 'a necessary complement to the speculative teaching on religion founded by Hegel'. Feuerbach had argued that in religion man projected his own attributes on to a divinity. Because of that, Engels concluded, 'Everything has changed' (2.196, 197, 198, 201, 237, 238).

Engels's campaign against Schelling concluded with another anonymous pamphlet, this one purportedly written by a pietist preacher of the sort Engels knew all too well from his days in Wuppertal. Engels's pietist praised Schelling for attacking philosophy and cutting away its ground, reason, from under its feet. The 'worldly wise' – Young Hegelians, no doubt – 'criticise the word of God with that corrupt reason . . . so as to make themselves God in His place'. Schelling was praised for his criticism of 'the notorious Hegel', who 'had such a pride in reason that he expressly declared it to be God when he saw that with it he could not come to another true God, higher than man'. Schelling, said Engels's pietist, 'has brought back the good old times when reason surrenders to faith'. In Berlin there were 'men of the world', philosophers, scholars, 'shallow un-Christian writers', and hypocrites who 'interfere most loudly in the government instead of leaving unto the King what is the King's'. These 'seducers . . . roam about in Germany and want to sneak in everywhere' (2.248, 250, 258–60, 264). A highly satisfactory battle in the press ensued.

Engels's subsequent articles were for opposition journals in Cologne, Leipzig and abroad in Switzerland. He had developed from a liberal journalist into a liberal, and in the Prussia of King Frederick William IV this made him a revolutionary. 'Prussian public opinion', he wrote, 'is centring more and more round two questions: representative government and, especially, freedom of the press', classic liberal demands (2.367). Concerning the latter, Engels quoted in one of his articles Section 151 of the Prussian Penal Code, which forbade 'insolent, disrespectful criticism and mockery of the laws of the land and government edicts', and he declared that he was 'honest enough to say straight out that I have every intention of provoking discontent and displeasure against Section 151 of the Prussian Penal Code'. He proposed 'to continue in the well-intentioned and decent fashion here indicated to awaken more than a little discontent and dissatisfaction with all obsolete and illiberal survivals in our state institutions' (2.305, 310–11). On the former issue – representative government – Engels remarked (with a weighty ellipsis) that 'Prussia's present situation closely resembles that of France before . . . but I refrain from any premature conclusions' (2.367).

3 Communist

The first visit Engels made to England was [in]
the summer of 1840. This was commemorated [in]
breathlessly romantic (Engels was twenty) y[oung]
remarks on the landscape between London and [...]
ever a land was made to be traversed by railways it [was]
(2.99). On his next trip, in late 1842, Engels's [...]
factual, eyewitness reporting were joined with a politic[al]
sciousness much deepened by the battles in Berlin. H[aving]
joined those at war with dogmatism, obscurantism, reaction a[nd]
orthodoxy, Engels brought the new revolutionary rationalism t[o]
bear on English life. This time he had a radical Cologne paper
at his disposal, and he set to work at once.

From London he attacked the 'ruling classes, whether
middle-class or aristocracy, whether Whigs or Tories' for their
blindness and obduracy, ever hostile to universal suffrage since
they would then be outvoted in the House of Commons by the
unpropertied. Chartism, the mass movement for liberal reform,
was 'quietly growing to formidable proportions', and Engels
wrote darkly of a débâcle for 'English Whiggery and Toryism'
(2.368–9).

While in Berlin in 1841–2 his political development reflected
that undergone by other Young Hegelians. After the relaxation
of press censorship their political views moved from a defence
of the rational state along more or less Hegelian lines to overt
criticism of Hegel, a rejection of middle-class liberalism, and
then advocacy of democracy, republicanism and social reform
to benefit the poor. 'Socialism' and 'communism' were at that
time employed interchangeably by many writers, though communists were assumed to be even more radical. One of the first
German communists was Moses Hess, who discussed communism at length with Engels when they met in Cologne, impart-
ing to him an optimistic doctrine of atheism and moral

ut declared himself a

ppertal' found com-
But the situation in
December 1842. In
ce before embark-
revolution, com-
nan were being
dern industry,
kt of social and
That particular
doctrine any-
rich Oswald'
conclusions by himself,
the only way of proceeding beyond
marks. But Engels was persuaded, and he used
analytic and journalistic gifts to support and enliven the
abstractions he found so convincing and exciting.

a brief excursion in
d in print in some
t characteristic
iverpool: 'If
is England'
gifts for
l con-
ving
nd

'The Internal Crises' offered an argument of great specificity and plausibility, and was in effect the theoretical prelude to Engels's masterpiece of 1844, *The Condition of the Working Class in England*. Engels boldly inquired if a revolution in England was possible or probable. 'Put it to an Englishman', he said, and you will get 'a thousand excellent reasons to prove that there can be no question at all of a revolution' – England's wealth, industry, institutions, flexible constitution, the fact that every disturbance of public order can only bring unemployment and starvation. But in taking this view the Englishman 'forgets the basis because of the surface appearance'. Engels then launched an economic analysis of industrial England: a country dependent on trade, and forced constantly to increase industrial output. Protective tariffs pushed up the price of English goods and the level of English wages; free trade would mean a disastrous flood of imports and the destruction of English industry. English markets were falling to the Germans and French. The 'contradiction inherent in the concept of the industrial state' was thus revealed philosophically and by his direct observation. The slightest fall in trade would deprive a considerable part of the working class of its bread; a large-scale trade crisis would leave the whole class without any at all. Almost half of the English population belonged to a class of 'unpropertied, absolutely poor people, a class which lives from hand to mouth, which multiplies rapidly'. The recent alliance of unorganised strikers with the Chartists in the riots of 1842 was founded on an illusion – revolution by legal means. The 'dispossessed', Engels claimed, without citing any evidence, had gained something useful: a realisation that 'only a forcible abolition of the existing unnatural conditions' could improve their circumstances. Though held back by respect for law they could not fail to cause a crisis, when fear of starvation would be stronger than fear of law. This revolution was 'inevitable', but it would be interests, not mere principles, that would carry it through. Principles could only develop from interests, and the revolution would be social, not merely political (2.370, 372, 373-4).

From Manchester, where his family had been partners for some years in a cotton-spinning business, Engels pursued his analysis of the proletariat by direct observation. Though English workers were better off, when employed, than the French or Germans, they still faced destitution with the 'slightest fluctuation in trade'. Their savings and mutual benefit funds dried up when unemployment became general, as Engels claimed was occurring in Glasgow: 'when English industry expands, there is always some region or other which suffers'. The state, he commented, did not care whether starvation was bitter or sweet; it locked these people up in prison or sent them to penal settlements, and when it released them it had 'the satisfaction of having converted people without work into people without morals'. When employed, Manchester workers endured a twelve-hour day. When unemployed, 'Who can blame them, if the men have recourse to robbery or burglary, the women to theft and prostitution?' (2.378, 379).

Engels's masterpiece of 1844 had three further preliminaries: articles written and published during 1843-4 on a broader theme – the social history of England. Engels took up Thomas Carlyle's recently published *Past and Present* by way of introducing this vast project, praising the author for his 'human point of view' but deploring 'vestiges of Tory romanticism' and his lack of acquaintance with German philosophy, so that all his views were 'ingenuous, intuitive'. Carlyle's complaints about the emptiness and hollowness of the age, his attacks on hypocrisy and lying, his criticism of competition and the economics of supply and demand were 'fair'. But he did not penetrate to the cause of these phenomena and hence did not discover the solution. There was in consequence 'not a syllable mentioning the English Socialists' (3.444, 461–2, 466).

In his next two articles Engels traced the English social revolution from eighteenth-century origins, particularly the development of the steam engine and mechanisation in textile and metal manufacture, listing inventions by Watt, Wedgwood, Hargreaves, Arkwright, Crompton and Cartwright, and noting improvements in communication by road, canal and rail. Yet these improvements benefited only the few, enslaved the many, and profoundly altered the values of English society.

This revolution through which British industry has passed is the foundation of every aspect of modern English life, the driving force behind all social development. Its first consequence was, as we have already indicated, the elevation of self-interest to a position of dominance over man. Self-interest seized the newly-created industrial powers and exploited them for its own purposes; these powers, which by right belong to mankind, became, owing to the influence of private property, the monopoly of a few rich capitalists and the means of the enslavement of the masses. Commerce absorbed industry into itself and thereby became omnipotent, it became the nexus of mankind; all personal and national intercourse was reduced to commercial intercourse, and – which amounts to the same thing – property, things, became master of the world. (3.469, 479–85)

The most important effect of this historical development, Engels wrote, was 'the creation of the proletariat by the industrial revolution'. Then he surveyed the English constitution and legal system, dismissing it as 'a jungle of lies and immorality', vastly out of touch with the new industrial society.

The *juste-milieu* esteem it a particular beauty of the English constitution that it has developed 'historically'; that means, in plain

German, that the old basis created by the revolution of 1688 has been preserved, and this foundation, as they call it, further built on. We shall soon see what characteristics the English constitution has acquired in consequence; for the moment a simple comparison of the Englishman of 1688 with the Englishman of 1844 will suffice to prove that an identical constitutional foundation for both of them is an absurdity and an impossibility. (3.487, 490, 512)

Promising to stick to 'empirical facts' rather than to the mythology of Blackstone, Magna Carta and the Reform Bill, Engels surveyed the monarchic, aristocratic and democratic elements of government. He concluded that the Crown and the House of Lords had lost their importance and that the House of Commons was all-powerful. The real question, he wrote, was: Who actually rules in England? His answer was, 'Property rules.' The middle class was dominant and the poor man had no rights; the constitution repudiated him and the law mistreated him. The 'struggle of democracy against the aristocracy in England' was a 'struggle of the poor against the rich' (3.492, 497, 513).

The battle for democracy, according to Engels, was but a transition to socialism. The battle of the poor against the rich could not be fought 'on a basis of democracy or indeed of politics as a whole'. Revolution would have to be *social* and move beyond political institutions to economic life and the reigning values in society. In his account of the development of English industrial society, Engels put Carlyle's complaints about cash-payment into the German philosophical context he said they lacked.

The abolition of feudal servitude has made 'cash-payment the sole relation of human beings'. Property, a natural, spiritless principle, as opposed to the human and spiritual principle, is thus enthroned, and ultimately, to complete this alienation, money – the alienated, empty abstraction of property – is made master of the world. Man has ceased to be the slave of men and has become the slave of *things*; the perversion of the human condition is complete . . . (3.476, 512–13)

According to Engels, his book *The Condition of the Working Class in England* was written 'from personal observation and authentic sources', and he challenged 'the English bourgeoisie' to prove him wrong in a 'single instance of any consequence'. The work was dedicated 'To the Working Classes of Great Britain', and the author's purpose was avowedly political. 'We Germans', he said, need to know the facts about England. While conditions in Germany were not in the 'classical form' found in England, the two countries had, at bottom, the same social order. The causes of proletarian misery and oppression in England were present in Germany, and the effects would eventually be the same. Official inquiries – Engels's major source of statistical information – into working-class life in England were thus critically relevant for 'German socialism and communism', the movement he aimed to further in this German-language work. It was researched in England, written up in late 1844 and early 1845, and published almost immediately (4.295, 297, 303). The book attracted considerable critical notice and was reprinted in 1848; a second German edition appeared in 1892, late in Engels's lifetime; and an English translation was published in New York in 1887. Engels was twenty-five when his first major work was published, and he had no lack of readers and critics.

Engels's book was biased and politically partial. His moral and philosophical position was clear throughout the work, and it presented a wholly unflattering account of the 'possessing class' and its role in a competitive economic system. He referred to his common cause with the English working class – the 'cause of Humanity' – and predicted that their wrath would erupt in a revolution beside which the French Revolution and the Terror would look like child's play. Hence it is not surprising that Engels's use of sources was highly selective. He chose reports, sometimes sensational ones from socialist newspapers, which emphasised the worst cases of poverty, degradation and suffering. From *The Times* and the *Northern Star* Engels extracted three particularly gruesome accounts. One concerned Ann Galway, aged forty-five, who lived at 3 White Lion Court, Bermondsey Street, London with her husband and nineteen-year-old son in a small room without bed or other furniture. Her dead body was reported, by the coroner for Surrey, to be 'starved and scarred from the bites of vermin'. 'Part of the floor of the room was torn up, and the hole used by the family as a privy.'

In a second case two boys appeared before the police magistrate in London, because 'they had stolen and immediately devoured a half-cooked calf's foot from a shop'. The mother of the two boys proved to be a widow living in dire poverty with her nine children.

When the policeman came to her, he found her with six of her children literally huddled together in a little back room, with no furniture but two old rush-bottomed chairs with the seats gone, a small table with two legs broken, a broken cup, and a small dish. On the hearth was scarcely a spark of fire, and in one corner lay as many old rags as would fill a woman's apron, which served the whole family as a bed . . . Her bedding she had pawned with the victualler for food.

A third case, concerning a widow who lived by charring, was similar. She and her sick daughter, aged twenty-six, lived in a back room no larger than a closet and had sold or pawned everything that they possessed.

In his defence Engels could only comment that he had cited the most horrifying cases deliberately – 'I know very well that ten are somewhat better off, where one is so totally trodden under foot by society' (4.301, 304, 323, 334–51).

But when Engels took his readers on 'wanderings' around Manchester, second-hand accounts receded and history, geography and sociology came to life. Engels's observations caught the complexity – in housing, industry, transport and sanitation – of life for the inhabitants of Manchester, and the differences in conditions for its citizens. Engels was, of course, a gentleman with access to the domain of the well-to-do, yet a communist who wanted more than a 'mere *abstract* knowledge of my subject' and so went with working-class companions into the poor sections of the town (4.297, 364). One of these was Mary Burns, an Irish mill-worker, who became his mistress and remained so until her death in 1863.

It was easy, Engels wrote, for successful Mancunians to avoid making such excursions themselves. The town itself was peculiarly built, 'so that a person may live in it for years, and go in and out daily without coming into contact with a working-people's quarter or even with workers'. By 'unconscious tacit agreement' and 'outspoken conscious determination' such districts were separated from the middle-class sections of the city – the central commercial district, deserted at night, and the outlying suburbs served by omnibuses. Between the two middle-class domains lay working-people's quarters, masked along the arterial roads by shop-fronts. Engels took Manchester to be a particularly pure result of private choice in industrial society.

I know very well that this hypocritical plan is more or less common to all great cities; I know, too, that the retail dealers are forced by the nature of their business to take possession of the great highways; I know that there are more good buildings than bad ones upon such streets everywhere, and that the value of the land is greater near them than in remoter districts; but at the same time I have never seen so systematic a shutting out of the working-class from the thoroughfares, so tender a concealment of everything which might affront the eye and the nerves of the bourgeoisie, as in Manchester. (4.347–8, 349)

Industry in Manchester adjoined the rivers and canals in working-class districts. In considering the historical geography of the area Engels was thoroughly analytic; his account of the Old Town was illustrated with 'a small section of the plan of Manchester' to characterise the 'irrational manner in which the entire district was built'.

Of the irregular cramming together of dwellings in ways which defy all rational plan, of the tangle in which they are crowded literally one upon the other, it is impossible to convey an idea. And it is not the buildings surviving from the old times of Manchester which are to blame for this; the confusion has only recently reached its height when every scrap of space left by the old way of building has been filled up and patched over until not a foot of land is left to be further occupied. (4.350–1)

Inadequate privies, polluted rivers, refuse and piggeries were all described, as were the 'one-roomed huts' and their inhabitants. All considerations of cleanliness and health were defied in the construction of this district. But old as it was, everything

arousing horror and indignation was of recent origin in the industrial epoch (4.353, 354–5).

What of the New Town? Pure accident determined the
grouping of houses in the Old Town, where spaces were called
courts, 'for want of another name'. But the orderly back-to-
back dwellings of the New Town produced bad ventilation *by
design*. Engels produced plans to show, as if from above, two
methods of constructing 'cottages' for working people, almost
always built by the dozen along streets *and* along the almost
invisible back alleys. Speculative building and leaseholding in-
teracted, according to Engels, even in the pattern of brickwork
construction; single rows of bricks laid end-to-end were used to
make cheap, flimsy outer walls (4.356, 357, 359).

Engels joined with earlier investigators of working-class
poverty in concluding that the 'working people of Manchester
and its environs live, almost all of them, in wretched, damp,
filthy cottages' in which 'no cleanliness, no convenience, and
consequently no comfortable family life is possible'. In such
dwellings 'only a physically degenerate race, robbed of all
humanity, degraded, reduced morally and physically to
bestiality, could feel comfortable and at home' (4.364).

Engels surveyed the poor state of working-class clothing,
food, tobacco, health, medicine, morals and working conditions
for adults and children, and dismissed the Factory Acts as inadequate and poorly enforced. Then his analysis moved to a
more general level and more sweeping conclusions.

Superimposed on this misery were the vagaries of the trade
cycle, arising from unregulated production and distribution,
carried on 'not directly for the sake of supplying needs, but for
profit, in the system under which everyone works for himself to
enrich himself'. The system was neither equal nor fair in its
effects: 'So the bourgeois certainly needs workers, not indeed
for his immediate living, for at need he could consume his
capital, but as we need an article of trade or a beast of burden –
as a means of profit' (4.378, 381).

Engels's account of working-class resistance to this state of
affairs was prescient yet over-logical. English workers 'cannot
feel happy in this condition' and 'must therefore strive to

escape'. The earliest and least fruitful form of this rebellion was that of crime. Machine-breaking and strikes were detailed, but the unions remained powerless against the great forces – competition and the trade cycle. The real importance of the unions, Engels concluded, was that they were 'the first attempt of the workers to abolish competition' among themselves *and* competition in the economic system as a whole. Engels reviewed the widespread occurrence of strikes and demonstrations and dismissed the Chartist and English socialist response as inadequate.

Hence it is evident that the working-men's movement is divided into two sections, the Chartists and the socialists. The Chartists are theoretically the more backward, the less developed, but they are genuine proletarians all over, the representatives of the class. The socialists are more far-seeing, propose practical remedies against distress, but, proceeding originally from the bourgeoisie, are for this reason unable to amalgamate completely with the working class. The union of socialism with Chartism, the reproduction of French communism in an English manner, will be the next step, and has already begun. Then only, when this has been achieved, will the working class be the true intellectual leader of England. Meanwhile, political and social development will proceed, and will foster this new party, this new departure of Chartism. (4.501, 502, 505, 507, 526–7)

From the evidence he had selected – because he thought it of the utmost significance – Engels predicted that the workers would come to perceive more clearly how competition affects them. They saw more clearly than the bourgeoisie that competition among capitalists causes commercial crises 'and that this kind of competition too, must be abolished' (4.508).

It was not to Engels's purpose to draw out evidence contrary to the cause he was promoting, for his account of the situation was not intended to be a mere reflection of circumstances, but was designed to assist certain developments in society and discourage others. While his work was avowedly partial to what he took to be working-class interests, critics today must think carefully before dismissing it for failing to be impartial, neutral and non-engaged. What would an impartial account of misery be like? Should one be neutral about suffering? What is the point of research and theorising if it does not help to alter the structure of an imperfect world?

Engels's prediction of a 'violent revolution, which cannot fail to take place' was not borne out (4.547). But working-class efforts to improve standards in the place of work and in housing, and to resist the hard effects of competition on individuals, have been influential in restructuring society more or less peacefully, a process to which Engels contributed by challenging reformers and their critics to look more closely at the plight of workers in a rapidly industrialising society.

4 Revolutionary

Engels's first meeting with Karl Marx was not a success. On his way to England in November 1842 Engels visited (for the second time that year) the offices of the radical Cologne paper which had been publishing some of his articles. Marx had been made editor in mid-October and had taken a no-nonsense line on contributions from the Berlin group of Young Hegelians, with whom Engels was associated.

As a scholar, philosopher and intellectual Marx was far ahead of Engels. At Trier, where he was born on 5 May 1818, two years before Engels, he was educated in the Latin, Greek and French classics at home, at school and at the house of his future father-in-law, Baron von Westphalen. Marx's parents were Jews converted to Lutheranism for political reasons, but neither Judaism nor Christianity played a major role in Marx's up-bringing, in contrast with the oppressive pietism experienced by Engels. In religion and politics Trier was a very much more liberal environment than Barmen; Marx imbibed the ideals of the French Revolution rather than the conservatism of the Prussian monarchy. Unlike Engels, Marx was a full-time uni-versity student, first at Bonn and then at Berlin, where he re-sisted (successfully) his father's attempts to see that he studied law. He pursued courses in philosophy and history and a more informal education among Young Hegelians in Berlin before Engels arrived. Marx's plans for an academic career were cut short, despite his completion of a doctoral thesis on Greek philosophy (accepted by correspondence at the University of Jena). When radicals were excluded from university employ-ment in Germany, he sought other means of developing the ideas current among Young Hegelians and of earning a living.

But Marx had only a fraction of the experience in journalism that Engels had had. Marx's only published efforts – three ar-ticles – had appeared in the Cologne paper: one on freedom of the press and two on historical and religious justifications for what he took to be illiberal absurdities in political life. Two of his projects as editor continued this approach: his criticism of feudal laws on wood-gathering and his exposé of poverty in the Moselle valley. After the first of these was published he broke decisively with the Berlin group, writing to Arnold Ruge in late November 1842 (just when Engels arrived) that 'to smuggle communist and socialist doctrines' into theatre reviews was 'inappropriate, indeed even immoral'. He utterly rejected 'heaps of scribblings, pregnant with revolutionising the world and empty of ideas, written in a slovenly style and seasoned with a little atheism and communism (which these gentlemen have never studied)' (1.393–4).

Why then, when Engels visited Marx in Paris two years later in August 1844, did he receive such a friendly welcome and an immediate proposal to collaborate on a pamphlet?

While in Manchester Engels had written an article in October–November 1842 entitled 'Outlines of a Critique of Political Economy', which was published in February 1844 in a journal co-edited by Marx and Ruge. Marx took notes, dating from early 1844, on the article (3.375–6), and much later in life described it as a 'brilliant sketch on the criticism of the economic categories' (1.330). That article had pride of place in Marx's account of his relationship with Engels. Engels's critical review of political economy (the economic theory of his day) must have impressed Marx, who was investigating the practical effects of the system of private property legalised and defended by the Prussian state. Marx was also well equipped to criticise Hegel's *Philosophy of Right*, the leading theoretical attempt to deal with private property and government, but he knew little of the French and English economists beyond what Hegel had put to his own uses. A critical study of political economy itself was clearly the next step for Marx in the serious consideration of the socially and politically disadvantaged in Germany and elsewhere in Europe.

Engels's advance into political economy arose from his concern with the social history of England, particularly the industrial revolution in the eighteenth century and earlier nineteenth century. Adam Smith, David Ricardo and James Mill were the

classic theorists of the virtues of private property and competition. As a communist, Engels proposed the abolition of both.
Marx's objection to communism (represented by the Berlin
group, among others) was not to its conclusions as such but to
the lack of real research and convincing argument to back them
up. Engels's essay was, at last, a communist work worth
reading.

Engels took political economy to be a science of enrichment
which developed as a result of the expansion of trade. 'The only
positive advance', he wrote, 'which liberal economics has made
is the elaboration of the laws of private property.' In his article
he attacked political economy as yet another manifestation of
the hypocrisy of the possessing class – the theme of the 'Letters
from Wuppertal' and other works of the preceding four years.
But within the hypocritical practice of competition he saw the
way to 'the great transformation to which the century is moving
– the reconciliation of mankind with nature and with itself'
(3.418, 421, 424).

As a moral critique of political economy Engels's work was
thorough and trenchant. Trade, like robbery, was based on the
law of the strong, and the envy and greed of merchants bore 'on
its brow the mark of the most detestable selfishness'. Claims
that trade was a bond of friendship among nations and individuals were but sham philanthropy, and the premises of
competition reasserted themselves soon enough.

The immediate consequence of private property is *trade* – exchange
of reciprocal requirements – buying and selling. This trade, like
every activity, must under the dominion of private property become
a direct source of gain for the trader; i.e., each must seek to sell as
dear as possible and buy as cheap as possible. In every purchase
and sale, therefore, two men with diametrically opposed interests
confront each other. The confrontation is decidedly antagonistic, for
each knows the intentions of the other – knows that they are
opposed to his own. Therefore, the first consequence is mutual mistrust, on the one hand, and the justification of this mistrust – the
application of immoral means to attain an immoral end – on the
other. Thus, the first maxim in trade is secretiveness – the concealment of everything which might reduce the value of the article in

question. The result is that in trade it is permitted to take the utmost advantage of the ignorance, the trust, of the opposing party, and likewise to impute qualities to one's commodity which it does not possess. In a word, trade is legalised fraud. Any merchant who wants to give truth its due can bear me witness that actual practice conforms with this theory. (3.418, 419–20, 422)

In modern times the liberal economic system had led to horrific results in factories, which dissolved common interests even in the family.

It is a common practice for children, as soon as they are capable of work (i.e., as soon as they reach the age of nine), to spend their wages themselves, to look upon their parental home as a mere boarding-house, and hand over to their parents a fixed amount for food and lodging. How can it be otherwise? What else can result from the separation of interests, such as forms the basis of the free-trade system? Once a principle is set in motion, it works by its own impetus through all its consequences, whether the economists like it or not. (3.423–4)

Pushing his analysis further Engels wrote that the 'law of competition is that demand and supply always strive to complement each other, and therefore never do so'. 'What are we to think', he asked, 'of a law which can only assert itself through periodic upheavals', namely the trading cycle of regular crises? It was ' a natural law based on the unconsciousness of the participants' (3.433–4).

Thanks to political economy and in particular the Malthusian theory of production and population, our attention had been drawn to the productivity of the earth and of mankind. Engels derived from this 'the most powerful economic arguments for a social transformation'. Private property had turned man into a commodity. Competition had 'penetrated all the relationships of our life and completed the reciprocal bondage in which men now hold themselves.' All this would drive us to 'the abolition of this degradation of mankind through the abolition of private property, competition and opposing interests'. Then, if production were carried on consciously, if producers knew how much consumers required, if

they were to share these products out, the 'fluctuations of competition and its tendency to crisis would be impossible' (3.434,
439–40, 442).

These themes, sketched by Engels, were taken up by Marx in
his 'Economic and Philosophical Manuscripts' (and the 'Notes
on James Mill'), begun in the spring of 1844. When Engels
himself arrived, Marx had a sizeable store of material towards
his own critical study of political economy, and some inkling
of what a large and difficult project it would be if it were done
thoroughly. There was all the more reason, then, for Marx to
interrupt his masterwork to deal urgently with his political
opponents. Engels seems to have agreed to a proposal made by
Marx that they dispose of the Young Hegelians altogether.
How better for Marx to do it than to enlist the services of a
reformed member of the Berlin circle?

Engels was also a writer with much more reputation than
Marx. Up to the time of this proposed joint project Marx had
published about two dozen articles in a bare handful of journals
and newspapers, some of which he had edited himself. Though
the Cologne paper (and its editor) achieved notoriety in late
1842 and early 1843, it was quickly suppressed. Such fame,
however, as Marx had achieved, did not derive so much from
the content of his own articles, as from the mixture of radical
and revolutionary sentiments expressed in the periodical as a
whole under his editorship. In a letter to Marx, Engels inquired
why his own name had been placed first on the title page of
their joint work, *The Holy Family: Critique of Critical Criticism*, since he had written so little of it (xxvii.22). He need
hardly have asked.

The Holy Family was not a fully collaborative publication,
in any case, since the chapters and even some sub-sections were
separately signed. The foreword identified the work as a polemic
preliminary to independent works in which 'we – each of us for
himself, of course – shall present our positive view': these were
Marx's critical works on political economy (and further
critiques of law, history, morals etc.) and Engels's *The Condition of the Working Class in England* (in press), and his projected social history of England. Now that Engels had moved on

from philosophy and liberalism to socialism and economics, he had only scorn for Young Hegelian criticism.

Criticism does nothing but 'construct formulae out of the categories of what exists', namely, out of the existing *Hegelian* philosophy and the existing social aspirations. Formulae, nothing but formulae. And despite all its invective against dogmatism, it condemns itself to dogmatism and even to *feminine* dogmatism. It is and remains an old woman – faded, widowed *Hegelian* philosophy which paints and adorns its body, shrivelled into the most repulsive abstraction, and ogles all over Germany in search of a wooer. (4.8, 20)

Engels followed Marx to Brussels in 1845 and on his own admission settled down to a junior role in the partnership, sensing, most probably, Marx's superior powers of analysis and unrelenting thoroughness. The two travelled to England in the summer of 1845 and visited Manchester, finding on their return to Brussels a reply to *The Holy Family* from the Young Hegelians. A clear statement of socialist premises was needed; previous criticism, as in *The Holy Family*, had proceeded from assumptions not fully articulated, and certainly not expressed in detail. Marx and Engels then embarked on what appears to be a genuinely collaborative work, *The German Ideology*, intended to settle their differences with the superficial but troublesome latter-day Hegelians, and to assist the authors in achieving 'self-clarification' (1.364).

The manuscript of *The German Ideology* was almost wholly in Engels's hand, with corrections and alterations by both authors. The pages were in fact divided into two columns, text on the left, additions on the right. Marx's handwriting was almost illegible, and it has been assumed that Engels was assigned the role of scribe in setting down a text composed aloud together. Philosophically the work resembled Marx's previous efforts more than Engels's, and the first section flowed directly from his 'Theses on Feuerbach', written early in 1845 before Engels's arrival in Brussels. In those few lines Marx launched an attack on 'all previous materialism' (5.3). When Engels looked back in 1888 after Marx's death he acknowledged that the eleven theses on Feuerbach were 'the first document in which is deposited the brilliant germ of the new world outlook' (II.359). And in another work of same year he wrote, echoing Marx's own words of 1859: 'How far I had independently progressed towards [Marx's premises] is best shown by my *The Condition of the Working Class in England*.' But 'when I again met Marx at Brussels . . . he had it already worked out' (I.29).

In *The German Ideology* various Young Hegelians were excoriated for their fantasy: that the 'relationship of men, all their doings, their fetters and their limitations are products of their consciousness'. By fighting phrases with phrases they were 'in no way combating the real existing world'. Marx and his collaborator took an opposing view: their premises were men, 'not in any fantastic isolation and fixity, but in their actual, empirically perceptible process of development', and their project 'the study of the actual life-process and the activity of the individuals of each epoch' (5.30, 37).

In the first part of *The German Ideology* there were passages which resembled very closely Engels's pre-Marxian works on the history of the industrial revolution and his remarks on the nature of communism. When Feuerbach, they wrote,

sees instead of healthy men a crowd of scrofulous, overworked and consumptive starvelings, he is compelled to take refuge in the 'higher perception' and in the ideal 'compensation in the species', and thus to relapse into idealism at the very point where the communist materialist sees the necessity, and at the same time the condition, of a transformation both of industry and of the social structure. (5.41)

When the state, law and property were discussed, Marx's early journalism and manuscripts were the likely source. The lengthy satires which took up the remainder of the book reflected work done by both men before *The Holy Family*, and in that work itself their *ad hominem* political critique was further developed. But the philosophical thread of *The German Ideology*, which ran throughout the text and gave it coherence, can be safely attributed to Marx, as Engels suggested. It should be stressed, however, that it was a philosophy opposed to mere philosophising, since it was intended to arise from real life and to restructure it in a practical, non-Utopian way.

Marx's 'Theses on Feuerbach' could hardly have come as a shock to Engels since his works of 1844 on social history and contemporary social conditions were so utterly compatible with them. 'All social life is essentially *practical*,' Marx wrote. 'All mysteries which lead theory to mysticism find their rational solution in human practice and in the comprehension of this practice' (5.5).

Engels, so far as we know, had now abandoned political economy to Marx, never enlarging on his 'brilliant' first critique, though there were traces of it in *The German Ideology*.

Or how does it happen that trade, which after all is nothing more than the exchange of products of various individuals and countries, rules the whole world through the relation of supply and demand – a relation which, as an English economist says, hovers over the earth like the fate of the ancients, and with invisible hand allots fortune and misfortune to men, sets up empires and wrecks empires, causes nations to rise and to disappear – whereas with the abolition of the basis, private property, with the communistic regulation of production . . . the power of the relation of supply and demand is dissolved into nothing, and men once more gain control of exchange, production and the way they behave to one another? (5.48)

Marx's first published work to show some of the fruits of his own economic reading and researches of 1845–6 was the *Poverty of Philosophy*, published in French in his name alone in 1847. Engels took on the task of journalist and publicist for the communist cause, now organised as the Brussels Correspondence Committee. The main aim, according to Marx, was to put German socialists in touch with English and French socialists (sc 28). Engels himself went to Paris to organise German workers, and he reported to the committee that he had obtained support (thirteen votes to two) for a definition of communism as (1) achieving the interests of the proletariat by (2) the abolition of private property and its replacement by the community of goods, achieving (1) and (2) by (3) the force of democratic revolution (sc 31–2). He was also emissary to London, and generally made it his business to see that communist groups adopted Marxian principles, rather than those of Young Hegelian or other provenance. The first congress of communists met in

London in the summer of 1847, where Engels drafted a 'Communist Confession of Faith' for discussion. Engels produced another version for Marx, the 'Principles of Communism', and later that year they both attended the second conference, held in London in late November and early December (6.96–103, 341–57). Marx and Engels were asked to use this draft and others in formulating a final document to which all communists could adhere.

Later in life Marx and Engels separately acknowledged that the thesis of the Communist Manifesto was Marx's own original contribution: that classes in society exist as a result of particular phases in the development of production, and that only the modern exploited class can accomplish the transition to a classless society (1.24–5, 28–9, 246; 11.344–5; SC 69–70). This was the thesis which arose from Marx's premises in *The German Ideology* and was actually elaborated there as well, though the elaboration, as I have suggested, bore in places considerable resemblance to Engels's early social history and comments on communism. But the early versions of the Manifesto were actually drafted by Engels alone, and elements of those drafts did appear at some length in the final document. Moreover the genre (something short and popular) and the purpose (gaining adherents among communists) were activities in which Engels was more directly involved than Marx, who was really more concerned, as he later put it, with finding the anatomy of bourgeois society in political economy.

Still, Marx took final responsibility for the text, since Engels left for Paris in early 1848 and Marx was pestered by the London communists for delivery late in January. In that sense the text was his (though it was, of course, unsigned). It was also intended to represent the views of a committee, and so there is reason to suppose that some sections of it reflected an attempt by Marx to meet or forestall objections lodged by others. He could probably have constructed it without the help of Engels's drafts or even perhaps without ever having met Engels at all, since the thesis and its elaboration flowed so logically from his works of 1842 and 1843. But Engels's early works, whose influence was acknowledged by Marx, must have made his progress through some of the elaboration in the Manifesto very much easier. In his early works, as we have seen, Engels adumbrated famous passages on the industrial revolution and the hypocrisy of the bourgeoisie, as well as the effect of competitive economic relations on women, children and the family. Moreover, the critical guide to English and continental socialism in part III of the Manifesto reflected his talent for writing brief accounts of contemporary philosophical debates.

At the very least Marx had important material presented to him by Engels which needed little alteration. Also, Engels's early works and his later collaboration in the composition of *The German Ideology* and the Manifesto may have been critical in swaying Marx from the baroque complexity displayed in his works of 1843 towards a more readable style. Whatever credit is assigned to Engels in the composition of the Manifesto, his efforts in securing the commission and the audience for it must be acknowledged. Translations into other European languages were made that year, and subsequently it was republished in German in 1872 with a new joint preface, and then in 1883 and 1890 with prefaces by Engels.

It is generally accepted that the Manifesto had no traceable influence on the revolutions of 1848. Marx and Engels, however, played an active though hardly world-historical role in those events. Marx edited a newspaper in Cologne; Engels contributed about eighty articles to it, and he wrote for other journals as well. Circulation of the Cologne paper is said to have been about 5–6,000 during its one-year existence. Marx aimed to assist democratic revolutions in Germany and elsewhere, following the outbreak of revolution in Paris in February 1848, and the policies urged by the paper reflected the programme announced in the manifesto. These were in part liberal measures which would command support against reactionary regimes and yet protect the interests of whatever working class was involved. Disappointment with the failure of middle-class revolutionaries in April 1849 led to the adoption of a more radical view of working-class political action shortly before the suppression of the paper in mid-May.

Engels himself was involved in unsuccessful revolutionary

scuffles in his home district of Elberfeld in May 1849, but the small crowd of insurgents was not joined by workers and militia in sufficient numbers to pose a mortal threat to the authorities. In June he joined revolutionary forces in south-west Germany on their unsuccessful campaign against the Prussians, wanting, so he said in a letter to Marx's wife, to save the reputation of his mentor's paper (SC 48). In late 1849 he made his way from Switzerland to Genoa and sailed to join the recently arrived Marx in exile in London.

5 Marxist

Engels was the first Marxist, and he had a defining influence on Marxism. A vast number of articles, pamphlets and reviews, and a respectable number of books occupied him from 1849 until his death in 1895. In many of these he attempted to explicate Marx's premises and views, to which he had substantially contributed. He also became Marx's reviewer and editor, writing prefaces for new editions of his (and their) works and preparing Marx's manuscripts for publication after the senior partner's death in 1883.

In Engels's first year in England he was wrapped up in the aftermath of the revolutionary events of 1848–9, very reasonably expecting them to continue after a period of apparent calm. Characteristically, Marx's first project at that time was to continue his political journal, now subtitled a political-economic review, promising 'a comprehensive and scientific investigation of the *economic* conditions which form the foundation of the whole political movement'. Engels's brief was evidently to contribute to a rather broader social and historical analysis in 'elucidating the period of revolution just experienced, the character of the conflicting parties, and the social conditions which determine the existence and struggle of these parties' (10.5). During the twelve months after November 1849 Engels published (for both English and German readers) his views on the revolutions of 1848–9 and the political controversy over the Ten Hours Bill for factory work in England. He also wrote a series of articles on 'The Peasant War in Germany', seeking to remind the German people of the 'clumsy yet powerful and tenacious figures of the Great Peasant War'. By looking at the events of 1525, he wrote, we 'shall see the classes and fractions of classes which everywhere betrayed 1848 and 1849 . . . though on a lower level of development' (10.399).

Engels's method in writing history was to draw from his sources the evidence required to demonstrate the truth of

Marx's view that the existence of classes in society depended on phases in the development of production and that only the modern proletariat was in a position to usher in the classless society. Engels sketched the history of German industry in the fourteenth and fifteenth centuries (cloth-making, metalwork, printing), and the political divergence of Germany from the pattern set elsewhere in Europe.

While in England and France the rise of commerce and industry had the effect of intertwining the interest of the entire country and thereby brought about political centralisation, Germany had not got any further than grouping interests by provinces, around merely local centres, which led to political division, a division that was soon made all the more final by Germany's exclusion from world commerce. In step with the disintegration of the *purely feudal* Empire, the bonds of imperial unity became completely dissolved, the major vassals of the Empire became almost independent sovereigns, and the cities of the Empire, on the one hand, and the knights of the Empire, on the other, began entering into alliances either against each other or against the princes or the Emperor. (10.401–2)

His summary of the German situation echoed the Communist Manifesto. 'The various estates of the Empire – princes, nobles, prelates, patricians, burghers, plebeians and peasants – formed an extremely confusing mass with their varied and highly conflicting needs.' And following the method of *The German Ideology* Engels unmasked theological controversies as in reality social and political. 'The revolutionary opposition to feudalism was alive throughout the Middle Ages. It took the shape of mysticism, open heresy, or armed insurrection, depending on the conditions of the time'. Using this framework Engels located three main camps: first, conservative Catholics – 'all the elements interested in maintaining the existing conditions, i.e. the imperial authorities, the ecclesiastical and a section of the lay princes, the richer nobility, the prelates and the city patricians'; secondly, Lutheran reform, which attracted moderates, i.e. 'the mass of the lesser nobility, the burghers, and even some of the lay princes who hoped to enrich themselves through confiscation of church estates'; and thirdly, peasants and plebeians, 'a *revolutionary* party whose demands and doc-

trines were most forcefully set out by [Thomas] Münzer'
(10.410, 413, 415–16).

Münzer was the character to whom Engels was most sympa-
thetic, but the lengthy account of the events of the peasant wars
of the 1520s was not by any means a simple paean to a hero of
the left. The two leaders, Luther and Münzer, truly 'reflected'
the attitudes of their parties. Luther's indecision corresponded
to the hesitant policies of the burghers; Münzer's revolutionary
energy corresponded to the most advanced plebeians and
peasants. Münzer, however, went far beyond their immediate
demands and in so doing found himself in an impossible situation. 'The worst thing that can befall the leader of an extreme
party is to assume power at a time when the movement is not
yet ripe for the domination of the class he represents.' Neither
Münzer's own movement nor the economic conditions in which
he found himself were ready for the social changes he en-
visaged: community of property, the equal obligation to work,
and the abolition of all authority. Such social changes as were
actually possible and under way were leading, Engels argued,
from feudalism to bourgeois society, a competitive commercial
system diametrically opposed to Münzer's notions. In this 'un-
solvable dilemma' what a leader '*can* do contradicts all his previous actions and principles and the immediate interests of his
party, and what he *ought* to do cannot be done' (10.427, 469–
70, 471).

Despite the analogies between the revolutions of 1848–9 and
1525 – divided and isolated revolutionary forces fighting opponents on the right and left – Engels predicted a successful out-
come for the modern revolutionary movement, because it was
not a domestic affair but an episode in a European event
(10.482).

Engels's study of the peasant war in Germany was the first
Marxist work of *history*. Engels demonstrated that in what
appear to be religious struggles all was not resolved in theological terms, and that behind 'a religious screen' lay the 'interests,
requirements and demands' of various classes. Similarly, he
argued that the revolution of 1789 in France was more than 'a
somewhat heated debate' on the advantages of constitutional

monarchy over absolutism, the July revolution of 1830 was not solely about the 'untenability of justice "by the grace of God" ', and the February revolution of 1848 was not simply 'an attempt at solving the problem: "Republic or monarchy?"' Behind these political struggles there were always the economic concerns of social classes (10.411–12). The methods and terms of Marxist historiography were largely set by Engels in this pioneering work.

The revolutionary optimism of 'The Peasant War in Germany', written in the summer and autumn of 1850, soon faded, and Engels withdrew from London for financial reasons, taking up a position with the family firm in Manchester as a clerk. By 1851 the communist leagues with which Engels was associated in his revolutionary years had collapsed. After that he had little time for formal political involvement because of his business career in Manchester, even when the Working Men's International Association (First International) was founded in 1864 to promote the cause of socialism. Among its founders was Marx himself, who devoted a considerable amount of his time to its congresses, committees and pronouncements. Engels drifted easily into the role of elder statesman and senior adviser, but not founder or organiser in the international socialist movement. Only after his retirement from the cotton-spinning works in 1869 could he take a seat on the General Council of the International in 1870 and bear some of the burden of correspondence with the increasing number of socialist parties and groups around the world. Engels naturally had a particular interest in the German Socialist Party founded in 1869, and after the eventual demise of the International in 1874 he played an active part as informal adviser to the highest ranks of the party leadership. His involvement with the Second International, founded in 1889, was at a similar remove, though one of his last public appearances was at its Zürich conference in 1893.

Not surprisingly Engels hated Manchester and the businessmen with whom he had to associate. On leaving the family firm in 1869 he moved swiftly to London to be near Marx. Mary Burns had died in 1863, and her sister Lizzie took her place in Engels's life until 1877, when Engels married her the day before she died. Engels's household was then managed by Lizzie's niece, and from 1883 by Helene Demuth, Marx's former house-keeper. After Helene's death in 1890 Louise Kautsky, the divorced wife of the German socialist Karl Kautsky, became Engels's secretary and housekeeper, and on her marriage to Dr Ludwig Freyberger in 1894 a physician joined the establish-ment in Regent's Park Road.

Engels was as generous with his time and money as with his advice. His beneficence to Marx and his immediate family saved them many times from fates more horrible than the poverty and misery they endured; by 1870 Engels was able to provide them with a measure of financial independence at the same time as providing for his own. Many other *émigrés* and visiting socialists benefited from his hospitality and assistance, and the surviving Marx children and grandchildren shared in Engels's substantial estate after he died of cancer of the throat on 5 August 1895.

In the early years of exile Engels also helped Marx by writing articles for him in English. Marx was asked to be a correspon-dent for the New York *Daily Tribune*, but could not, until 1853, manage English well enough to write articles on his own. Engels acted as author and translator, and Marx received the fees. The series 'Revolution and Counter-Revolution in Ger-many', written in 1851–2, and republished twice in 1896 (in English and in German translation), was attributed to Marx rather than Engels until their correspondence on the matter was published in 1913. In this work Engels considered at length the revolutionary events in Germany in 1848–9 which he had wit-nessed and chronicled in print at the time – just three years earlier, or less. Marx undertook a similar task in his series of articles, 'The Class Struggles in France' (written in the first half of 1850), and in the continuation of the story in 'The Eighteenth Brumaire of Louis Bonaparte' (written in very late 1851 and early 1852). Engels, however, never produced the con-tinuation to his series in which he had promised to 'throw a parting glance upon the victorious members of the counter-revolutionary alliance', as Marx had done in 'The Eighteenth Brumaire' for France (11.96).

In 'Revolution and Counter-Revolution in Germany' Engels pursued the Marxist programme with respect to current history and future political developments. His task was to explain the principal events and to give clues to the direction which the next and perhaps not so distant revolutionary outbreak would take in Germany. Just as the revolutionary struggles of medieval times were not, despite appearances, to be explained in terms of theological disagreements, so the causes of the outbreak and defeat of the recent revolution were not, despite appearances, to be sought in the accidental efforts or treacheries of some of the leaders, but in the social state and 'conditions of existence' of each of the nations. Engels covered the economic, sociological, political, literary and philosophical background to the insurrections of 1848 in Vienna and Berlin, and he concluded that after victory the 'liberal bourgeoisie . . . turned round immediately' upon its working-class allies – the 'popular and more advanced parties' – and made 'an alliance with the conquered feudal and bureaucratic interests' (11.6, 7, 39).

This incomplete revolution found its epitome, according to Engels, in the German National Assembly at Frankfurt am Main; it engaged in 'a factitious, busybody sort of activity, the sheer impotence of which, coupled with its high pretensions, could not but excite pity and ridicule'. All those events were dependent on the fortunes of revolutionary struggles in France, first in February with the proclamation of the republic, and then the decisive action in June 1848. 'It could be fought in France only,' Engels wrote, 'for France, as long as England took no part in the revolutionary strife, or as Germany remained divided, was, by its national independence, civilisation and centralisation, the only country to impart the impulse of a mighty convulsion to the surrounding countries'. The June defeat of the working people by other classes, which were supported by the army, was crucial. For Engels it was 'evident to everyone that this was the great decisive battle which would, if the insurrection were victorious, deluge the whole continent with renewed revolutions, or, if it were suppressed, bring about an, at least momentary, restoration of counter-revolutionary rule' (11.31–2, 51, 92).

In Engels's analysis the task of mid-nineteenth-century German revolutionaries was daunting, and heavily influenced by events in the more advanced countries of England and France. His diagnosis of the German situation revealed it to be disappointingly similar in 1850 to what had been a fresh tragedy in 1525.

The preceding short sketch of the most important of the classes, which in their aggregate formed the German nation at the outbreak of the recent movements, will already be sufficient to explain a great part of the incoherence, incongruence and apparent contradiction which prevailed in that movement. When interests so varied, so conflicting, so strangely crossing each other, are brought into violent collision; when these contending interests in every district, every province are mixed in different proportions; when, above all, there is no great centre in the country, no London, no Paris, the decisions of which, by their weight, may supersede the necessity of fighting out the same quarrel over and over again in every single locality; what else is to be expected but that the contest will dissolve itself into a mass of unconnected struggles, in which an enormous quantity of blood, energy and capital is spent, but which for all that remain without any decisive results? (11.12)

Marx's economic studies, undertaken again in earnest in the 1850s, provided an optimistic dimension to Engels's political life that recent events and *émigré* politics could not, since Marx believed he was demonstrating that the capitalist system could not last much longer. A European and indeed world capitalist crisis was, in their view, the foundation of revolutionary progress.

After a decade of poverty, illness and the distractions of journalism, the first published instalment of Marx's masterpiece, a critique of political economy, was published in 1859. This (rather livelier) version of what was to become, among other things, the first chapters of the first volume of *Capital* appeared in German as *A Contribution to A Critique of Political Economy.* Marx hatched a plan whereby Engels was to review it, asking him in a letter of 19 July 1859 for something brief on the method and what was new in its contents. Nervously, Marx prompted Engels on 22 July with further suggestions (xxix.460, 463). A two-part review duly appeared (a

promised third section on the actual economic content of the
work significantly never materialised), and Engels became at
once the first populariser of Marx's innovations in social science
and the first commentator on his critical method. This work has
been more influential than Marx's later attempt to popularise
his economic material as lectures in 'Wages, Price and Profit',
delivered in 1865. His own comments on his method – the very
brief ones published in his lifetime, such as the 1872 afterword
to *Capital*, and the lengthier assessments posthumously published from his manuscripts, such as the 'Introduction' of 1857
which opens the *Grundrisse* notebooks – have only recently
come to the fore.

Following once again the method of *The German Ideology*
and the Manifesto, Engels approached Marx's achievements by
way of German economic history – the failure, after the
Reformation and the peasant wars, to develop the bourgeois
conditions of production visible in Holland, England and
France. The science of political economy in Germany consequently made little progress, and contemporary German writing
on the subject was dismissed by Engels as 'a mush consisting of
all sorts of extraneous matter, with a spattering of eclectic-
economic sauce, such as would be useful knowledge for a state-
employed law school graduate preparing for his final state board
examination'. When the German proletarian party appeared on
the scene (in the 1840s), scientific German economics was born.
The new economics, he wrote, was 'grounded essentially upon
the *materialist conception of history*' applicable to 'all historical
sciences'. In 'our materialist thesis', wrote Engels, 'it is demonstrated in each particular case how every time the action origi-
nated from direct material impulses, and not from the phrases
that accompanied the action' (1.366, 367, 368, 369). Engels's
phrase 'the materialist conception of history' brought Marxism
into existence.

In the second part of his review Engels added a further important element to this basic outlook: Marx's 'dialectical
method'. To explain it he employed three distinctions. In the
first Engels contrasted Hegel's achievements in handling 'inter-
connection' and 'categories' in science with the 'old meta-

physical manner of thinking', which he identified with the use of 'fixed categories' that reflected a 'new natural-scientific materialism . . . almost indistinguishable theoretically from that of the eighteenth century'. He then associated this metaphysics with 'bourgeois workaday understanding' which 'stops dead in confusion' when confronted with the separation of 'essence from appearance, cause from effect'. Despite Hegel's achievements in relating the development of thought to world history, the great philosopher had produced a dialectic in which 'the real relation was inverted and stood on its head'. It was 'abstract and idealist'. Only Marx was equipped to 'undertake the work of extracting from the Hegelian logic the kernel which comprises Hegel's real discoveries' (1.370, 371, 372, 373).

Marx's 'dialectical method', in which the 'idealistic trappings' were removed from Hegelian logic, constituted the second distinction introduced by Engels in his review, though he did not specify the 'simple shape' in which Marx's 'dialectic' became 'the only true form of development of thought' (1.373).

In the third distinction Engels attempted to contrast the 'historical' with the 'logical' method within Marx's 'dialectical' criticism of economics. Sweepingly Engels declared that historical events *and* their 'literary reflection', e.g. in economic theory, proceeded 'from the most simple to the more complex relations'. This development he then identified with the 'logical development' of 'economic categories'. The logical method in criticising political economy was therefore merely a paring down of historical 'leaps and zigzags', the exclusion of material 'of minor importance', and the omission of the full history of 'bourgeois society'. The 'logical method' was thus a 'reflection of the historical course in abstract and theoretically consistent form' (1.373).

In this logical analysis each economic relation, according to Engels, had 'two sides'. Each was considered by itself, and then their interaction. 'Contradictions will result which demand a solution' in the 'real process', not merely in 'an abstract process of thought'. Solutions, he wrote, had been brought about 'by the establishment of a new relation whose two opposite sides we shall now have to develop, and so on' (1.374).

At the opening of his review Engels quoted extensively from Marx's preface to *A Contribution to A Critique of Political Economy*, in which Marx himself outlined the 'guiding thread' of his studies. Engels's formulation of 'the materialist conception of history' and his three methodological distinctions followed, by way of explanation. Engels's account set the terms of interpretation for Marx's text and established the terms for the explication of Marxism itself.

Whether or not Engels's interpretation was correct, it was undoubtedly a gloss. Where Marx had written of his 'guiding thread', Engels wrote of 'the materialist conception of history'. In Marx's account it was Hegel's work on political economy in the *Philosophy of Right* which attracted his attention in resolving doubts about 'so-called material interests' and 'economic questions'; Engels introduced into this context the explicit consideration of metaphysics, materialism, idealism, dialectic, interaction, contradiction and reflection, as quoted above. Where Marx resolved to ascend 'from the particular to the general' in his consideration of capital, Engels espoused a vast thesis concerning the history and intellectual development of western society. (1.361, 362). While Marx was fully attuned to Hegel's critique of traditional logic and yet to the alleged error of Hegel's idealist premises (taking ideas to be the stuff of reality), he could not be said to have employed the method outlined by Engels. Engels's account did not convey the investigative armoury at Marx's disposal in his work on political economy, since Marx used a multiplicity of techniques and distinctions as he found them appropriate. Engels's bare schema of consideration, contradiction and solution gave the impression that Marx was merely reflecting a historical course, rather than subjecting a body of economic theory to logical, philosophical, mathematical, sociological, political and historical analysis. Engels's concepts of materialism, metaphysics, dialectic, interaction, contradiction and reflection reappeared, with a good deal more specificity, in his later writings, and I shall consider those concepts in chapters 6 and 7.

When the first volume of *Capital* was published in Hamburg in 1867 Engels again reviewed Marx's work anonymously, this

time in no less than seven different German and English period-
icals, helping to counter the critical silence which characteristically greeted Marx's published works. Two other reviews by
Engels were drafted but not actually published. Depending on
his assessment of the readership of each journal Engels recommended Marx's work for different reasons, but most readers
were given to understand that an important contribution to the
science of political economy was under review, a work far sur-
passing anything previously accomplished. Its supreme achievements lay in explaining the origin of profit – a puzzle to all
previous political economists, according to Engels – in terms of
the newly introduced categories of surplus value, surplus
labour, and the purchase of labour power. Marx's technique in
treating economic relations was described as 'a wholly new,
materialistic, natural-historical method', and his analysis of the
'law' of historical development was compared with the work of
Darwin and the whole of modern geology. Some readers received in addition an explicit account of the political deductions
drawn by Marx in the course of his critique of political
economy.

These, strictly scientifically proved – and the official economists
take great care not to make even an attempt at a refutation – are
some of the chief laws of the modern, capitalist, social system. But
does this tell the whole story? By no means. Marx sharply stresses
the bad sides of capitalist production but with equal emphasis
clearly proves that this social form was necessary to develop the
productive forces of society to a level which will make possible an
equal development worthy of human beings for *all* members of
society. All earlier forms of society were too poor for this. Capitalist
production is the first to create the wealth and the productive forces
necessary for this, but at the same time it also creates, in the
numerous and oppressed workers, the social class which is compelled
more and more to take possession of this wealth and these productive forces in order to utilise them for the whole of society . . .
(1.462, 463, 464, 468–9; xvi.217)

By 1878 Engels had also become, in a small way, Marx's
biographer, contributing a sketch for a German almanac on
'the man who was the first to give socialism, and thereby the

whole labour movement of our day, a scientific foundation'. He chose to dwell on only two of Marx's discoveries: his 'new conception of history' and the 'final elucidation of the relation between capital and labour'. The discussion of the former proceeded in very positive terms. 'Marx has proved that the whole of previous history is a history of class struggles', and that 'these classes owe their origin and continued existence' to the 'particular material, physically sensible conditions in which society at a given period produces and exchanges its means of subsistence'. From this point of view 'all the historical phenomena are explicable in the simplest possible way – with sufficient knowledge of the particular economic condition of society'. In the published version of Engels's rather more famous 'Speech at Marx's Graveside', this point of view was again likened to Darwin's work – described as 'the law of development of organic nature' – and was termed 'the *law* of development of human history'. The theory of surplus value became, in Engels's eulogy, 'the *special law* of motion governing the present day capitalist mode of production'. He then alluded rather vaguely to other 'independent discoveries' made by Marx and linked the man of science with the 'revolutionist' (II.156, 162–6, 167; emphases added).

Science was for Marx a historically dynamic, revolutionary force. However great the joy with which he welcomed a new discovery in some theoretical science whose practical application perhaps it was as yet quite impossible to envisage, he experienced quite another kind of joy when the discovery involved immediate revolutionary changes in industry, and in historical development in general. (II.168)

In the years after Marx's death in 1883 Engels produced prefaces to new editions of their *Communist Manifesto* (three editions), of his own *The Condition of the Working Class in England* (two editions), and of six works by Marx, the *Poverty of Philosophy*, *The Eighteenth Brumaire of Louis Bonaparte*, *The Civil War in France*, *Wage-Labour and Capital*, *The Communist Trial in Cologne* and *The Class Struggles in France*. To these works he contributed editorial notes and changes, but his principal projects as Marx's editor were the second and third volumes of *Capital* (with prefaces), derived from Marx's unpublished manuscripts.

Engels's role as guardian of what he took to be Marx's dis-coveries in historical and economic science can be illustrated by turning to two of the prefaces mentioned above. The republica-tion as a pamphlet in 1895 of Marx's articles *The Class Struggles in France* was introduced by Engels as 'Marx's first attempt to explain a section of contemporary history by means of his materialist conception, on the basis of the given economic situation'. Marx's task, according to Engels, was 'to demon-strate the inner causal connection' in a historical development which was for Europe both critical and typical. The object was 'to trace political events back to effects of what were, in the final analysis, economic causes'. However, the reader who looked in *The Class Struggles* for a precise list of economic causes would be disappointed, as Engels recognised in a qualifying passage. Economic factors, he wrote, were 'complicated and ever-changing', so 'the materialist method has here quite often to limit itself to tracing political conflicts back to the struggles between the interests of the existing social classes and fractions of classes'. Political parties can then be proved to be the political expression of those classes and fractions of classes (1.118–19).

Engels was actually suggesting a major source of error in Marx's account, written in 1850, since 'the economic history of a given period can never be obtained contemporaneously', but only after the consideration of, for example, statistical material which must be collected subsequently. Marx's work on the events of 1848–9, however, stood up to a double test, in Engels's view: a subsequent investigation of the economic circumstances of the period, and Marx's own reconsideration of 1848–9 in the light of Louis Bonaparte's *coup d'état* of late 1851. Still, Marx's account of contemporary political events in terms of classes, parties and individuals fitted rather poorly into Engels's meth-odological mould, with its emphasis on *'ultimate* economic causes' in 'the movement of industry and trade' (1.119, 120, 121).

When a series of Marx's articles of 1849 was republished in

1891 as *Wage-Labour and Capital* Engels considered in his introduction whether 'Marx himself would have approved of an unaltered reproduction of the original' as 'a propaganda pamphlet'. 'Marx', he wrote, 'would certainly have brought the old presentation dating from 1849 into harmony with his new point of view', elaborated in *A Contribution to A Critique of Political Economy* of 1859 and the first volume of *Capital*, published in 1867. Engels therefore warned his readers that 'this is not the pamphlet as Marx wrote it in 1849 but approximately as he would have written it in 1891'. All Engels's alterations turned on one point: the worker, according to the original text, sold his *labour* to the capitalist for wages, whereas in Marx's mature conception he sold his labour *power*. This apparently minor alteration enabled Marx to find the way out of a blind alley in economics, by formulating the theory of surplus value (1.70, 71, 75). But in his work of 1849 Marx had actually referred to labour as 'the creative power by which the labourer not only replaces what he consumes, but gives to accumulated labour a greater value than it previously possessed'. This was the substance, though not the crisper terminology, of the 1859 conception. Engels rectified this with enthusiasm (vi.409).

Up to the present day Engels's editing of the manuscript drafts left by Marx for the second and third volumes of *Capital* has not been scrutinised, because the manuscripts themselves, said to be in Moscow, have not been available. They are due to be published in the remaining decades of this century, and we will then know exactly how in these works Engels conceived 'the bounds of editing', his phrase in the preface to the third volume.

From the time that Marx's critique of political economy began to reach the press in 1859, Engels's views on Marx's work, his own work, history and politics became increasingly coloured by the language of ultimate causation and scientific laws of development. Those themes received an independent elaboration in the substantial works undertaken by Engels between 1870 and 1895 on his own account. Those were the works that provided – and still provide – millions of readers with the classic exposition of Marxism.

6 Scientist

The potential audience for Marxist ideas increased very dramatically in 1875 with the formation of a large, united and electorally successful socialist party in Germany. Engels took up the challenge.

Initially his approach was indirect – a critique of the works of Eugen von Dühring, an academic convert to socialism whose influence within the party was growing. In response to promptings from the anti-Dühring faction within the leadership Engels undertook the task of clarifying 'our position *vis-à-vis* this gentleman', as he put it in a letter to Marx of 24 May 1876 (xxxiv.12–13). In the 1870s Dühring had published a *Critical History of Political Economy and Socialism*, a *Course in Political Economy*, and a *Course in Philosophy as a Strictly Scientific World Outlook and Pattern for Life*. Engels logically took the *Course in Philosophy* as the major target for his attack, since it 'better exposes the weak sides and foundations of the arguments put forward in the *Economy*'. Dühring's 'banalities', he wrote to Marx, were revealed in a 'simpler form than in the economy'. The structure of Engels's polemic was largely dictated by Dühring's rambling synopsis of 'the philosophy of reality'. According to Engels, Dühring had produced precious little 'actual philosophy – formal logic, dialectics, metaphysics etc.' and had a laughable method, taking 'everything to be natural that seems natural' (SC 305; xxxiv.27).

Herr Eugen Dühring's Revolution in Science, generally known as *Anti-Dühring*, appeared in instalments in a German socialist newspaper in 1877–8, then in three pamphlets, and again as a book just before the censorship imposed in Germany by the anti-socialist law of 1878. The work caused a considerable stir within the socialist party. Three chapters were published as *Socialism: Utopian and Scientific* in a French translation in 1880, and in German in this form in 1883. The complete book reappeared in 1886 and 1892, and by the latter date *Socialism: Utopian and Scientific* was circulating, so Engels claimed, in ten languages. 'I am not aware', he wrote, 'that any other socialist work, not even our *Communist Manifesto* of 1848 or Marx's *Capital*, has been so often translated. In Germany it has had four editions of about 20,000 copies in all' (II.94–5). Engels had put Marxism on the map.

According to Engels there were three reasons for pursuing Dühring's writings. The first was 'to prevent a new occasion for sectarian splitting and confusion from developing within the party, which was still so young and had but just achieved definite unity'. Dühring's views were being accepted as socialist without proper qualification; certain persons were ready to spread this doctrine among the workers; and the editorial policy of the party paper was being subverted.

The second was what Engels called in 1878 'the opportunity of setting forth in a positive form my views on controversial issues which are today of quite general scientific or practical interest'. While his work did not represent an 'alternative' system, Engels hoped that 'the reader will not fail to observe the connection inherent in the various views which I have advanced'.

Thirdly Engels aimed to warn his readers against other German systems of 'sublime nonsense', in which 'people write on every subject which they have not studied, and put this forward as the only strictly scientific method'. Dühring was merely 'one of the most characteristic types' promoting 'bumptious pseudoscience'. Still, Engels admitted frankly to being a dilettante in jurisprudence and natural science, limiting himself in those subjects to 'correct, undisputed facts' (AD 9, 10, 11).

Gradually the second project – the publication of 'positive views' – overtook the other considerations in Engels's mind. In the 1885 preface to the second edition of *Anti-Dühring*, written some two years after Marx's death, Engels wrote that his polemic 'was transformed into a more or less connected exposition of the dialectical method and of the communist world-outlook fought for by Marx and myself'. 'This mode of outlook', he wrote, 'now finds recognition and support far beyond the boundaries of Europe, in every country which contains on the one

hand proletarians and on the other undaunted scientific theore-
ticians'. This public, according to Engels, was keen enough 'to
take into the bargain the polemic against the Dühring tenets
merely for the sake of the positive conceptions'. What had been
described in the 1878 preface as 'my views' became in Engels's
later accounts a matter of joint authorship (AD 13). And in the
1892 preface to *Socialism: Utopian and Scientific* Engels wrote
that to publicise the 'views held by Marx and myself' was the
'principal reason which made me undertake this otherwise un-
grateful task'. Dühring's systematic comprehensiveness gave
him the opportunity of developing these joint views on such a
great variety of subjects (II.94). Curiously the original introduction to the pamphlet was by Marx but signed with the name of
Paul Lafargue, the French socialist who was his son-in-law
(xix.181–2, 564).

By 1894 Marx loomed even larger in Engels's view of *Anti-
Dühring*, since Engels then added some of Marx's manuscript
material to chapter 10. Having previously cut Marx's drafts for
the section of the work on political economy, Engels incorpor-
ated what was cut and repeated his acknowledgement of 1885,
the first time he had revealed publicly that Marx had helped
him in composing a small part of *Anti-Dühring*.

In the opening chapter of the work as originally published
Engels enlarged on the distinctions of his 1859 review of
Marx's *A Contribution to A Critique of Political Economy* –
metaphysics and dialectic, idealism and materialism, and the
historical and logical approach to the development of capitalism.
Marx received credit for discovering (1) the 'materialistic conception of history', and (2) 'the secret of capitalist production
through surplus value'. About the former Engels commented:

it was seen that *all* past history was the history of class struggles;
that these warring classes of society are always the products of the
modes of production and exchange – in a word, of the *economic*
conditions of their time; that the economic structure of society
always furnishes the real basis, starting from which we can alone
work out the ultimate explanation of the whole superstructure of
juridical political institutions as well as of the religious, philosophical and other ideas of a given historical period.

And he summarised the second discovery – the theory of surplus value – as follows:

It was shown that the appropriation of unpaid labour is the basis of the capitalist mode of production and of the exploitation of the worker that occurs under it; that even if the capitalist buys the labour-power of his labourer at its full value as a commodity on the market, he yet extracts more value from it than he paid for; and that in the ultimate analysis this surplus value forms those sums of value from which are heaped up the constantly increasing masses of capital in the hands of the possessing classes. (AD 37, 38)

Engels was the father of dialectical and historical materialism, the philosophical and historiographical doctrines developed by late nineteenth- and early twentieth-century Marxists. Those doctrines became the basis of official philosophy and history in the Soviet Union and in most other countries that declare themselves Marxist. They have also been an important focus of debate within Marxist political groups in non-Marxist countries. The terms of those doctrines have become familiar in academic philosophy and historiography, chiefly through the works of writers unconnected with the Soviet Union. Engels developed his dialectical views as expressed in 1859 in the first (1878) edition of *Anti-Dühring*, though they were far from neatly formulated. In the opening chapter Engels disposed of 'metaphysics':

To the metaphysician, things and their mental reflexes, ideas, are isolated, are to be considered one after the other and apart from each other, are objects of investigation fixed, rigid, given once for all. He thinks in absolutely irreconcilable antitheses. 'His communication is "yea, yea; nay, nay"; for whatsoever is more than these cometh of evil.' For him a thing either exists or does not exist; a thing cannot at the same time be itself and something else. Positive and negative absolutely exclude one another; cause and effect stand in a rigid antithesis one to the other. (AD 31)

Claiming on the contrary that 'the two poles of an antithesis' actually interpenetrate, Engels wrote that dialectics, as opposed to 'metaphysics' (which overlooks this interpenetration), 'comprehends things and their representations in their essential connection, concatenation, motion, origin and ending', and that 'Nature is proof of dialectics.' Modern materialism, for that reason, embraced 'the most recent discoveries of natural science' and was 'essentially dialectic' (AD 32, 33, 35, 36). In later chapters Engels then considered 'quantity and quality' and 'negation of the negation', two other dialectical laws. 'Dialectics', he wrote, 'is nothing more than the science of the general laws of motion and development of nature, human society and thought' (AD 143, 168–9).

When considering the 'materialist conception of history' in *Anti-Dühring*, Engels linked this view with his 'dialectical' view of science by claiming that 'social forces work exactly like natural forces' and that 'the final causes of all social changes and political revolutions are to be sought . . . in changes in the modes of production and exchange . . . not in the *philosophy*, but in the *economics* of each particular epoch' (AD 316, 331). In the 1885 preface to *Anti-Dühring* Engels made an even more explicit connection between his views on dialectics and Marx's work on political economy and the development of modern industrial society: 'Marx and I were pretty well the only people to rescue conscious dialectics from German idealist philosophy and apply it in the materialist conception of nature and history.' Engels continued his theme, writing that he aimed to convince himself in detail 'of what in general I was not in doubt':

that in nature, amid the welter of innumerable changes, the same dialectical laws of motion force their way through as those which in history govern the apparent fortuitousness of events; the same laws as those which similarly form the thread running through the history of the development of human thought and gradually rise to consciousness in the mind of man. (AD 15, 16)

The distillation of Engels's dialectics contained in the 1885 preface put the text of *Anti-Dühring* sharply into focus, in contrast to the rambling work of 1878 known to Marx.

Engels's notion of metaphysics was unusual in that he defined it as a particular philosophical position (the beliefs that concepts have fixed referents and that truth and falsity are solely and unambiguously attributes of propositions), rather than as a bare framework of abstract, general concepts which might be filled out with substantive philosophical views of what there is and why. His account of dialectic as a process of development through contradiction (or antitheses or opposites) accorded with Hegel's efforts to specify whatever contradictions arose in the development of the phenomena he investigated. Both Hegel and Engels, however, tended to write as if the dialectic reflected a necessary, inevitable process of development to which human agency was ultimately subordinate or even subjected, and both Hegel and Engels considered natural processes to be *in themselves* dialectical, implying a kind of knowledge denied by most modern philosophers. Marx, by contrast, concluded from his economic and political account of capitalist society that revolution was, we might say, as good as inevitable, without invoking a notion of historical necessity. Similarly, he did not venture into the murky realm of a causal linkage between material phenomena and human behaviour beyond a notion that the material conditions of production create possibilities for human agency and at the same time set limits to what can be accomplished. In his afterword to the first volume of *Capital* he identified a rational dialectic as one which included in a positive understanding of a state of affairs an understanding of its negation. While Hegel hardly came nearer to defining the dialectic than his comment in the introduction to the *Science of Logic* that it was a grasping of the positive in the negative, Engels identified the dialectic with natural laws of motion in nature, motion in history (presumably the development of events), and motion in thought (presumably the rules of formal logic). The alleged linkage between matter in motion (which Engels studied through chemistry and physics) and history and thought was merely asserted and unsurprisingly never specified. However, neither Hegel nor Engels nor Marx, whatever they severally understood by the dialectic, was so jejune as to employ the triadic formula thesis–antithesis–synthesis that has been so often, and so mistakenly, attributed to them. Indeed Marx mocked outright this very approach to Hegelian philosophy. The triadic formula was invented by Heinrich Moritz Chalybäus, an early commentator on Hegel shortly after the master's death. This interpretation did not illuminate Hegelian thought – rather the reverse – and it had the further consequence that the method and content of works by Marx and Engels have been seriously misrepresented.

Engels had a positivist view of science – 'the accumulating facts of natural science compel us' to a recognition of 'the dialectical conception of nature'; and he had a determinist view of social science, searching for ultimate causation (AD 19). He was also a strict materialist. In a passage of 1875 or 1876 from the *Dialectics of Nature* (a work not published until 1927) he wrote that matter itself accounts for all causation and consciousness:

we have the certainty that matter remains externally the same in all its transformations, that none of its attributes can ever be lost, and therefore, also, that with the same iron necessity with which it will exterminate on the earth its highest creation, the thinking mind, it must somewhere else and at another time again produce it. (DN 54)

Engels in fact interrupted these scientific investigations to work on *Anti-Dühring*. The immediate impulse for Engels to take up a dialectical interpretation of natural science had been his highly critical reaction to the second edition of Ludwig Büchner's *Man and his Place in Nature in the Past, Present and Future. Or: Where did we come from? Who are we? Where are we going?* The plan for a critique arose very early in 1873, and in a letter to Marx of 30 May he set down his 'dialectical ideas on the natural sciences' and asked for help.

In bed this morning the following dialectical ideas on the natural sciences came into my head:
The subject-matter of natural science – matter in motion, bodies. Bodies cannot be separated from motion, their forms and kinds can only be known through motion; of bodies out of motion, out of relation to other bodies, nothing can be asserted. Only in motion does a body reveal what it is. Natural science therefore knows bodies by considering them in their relation to one another, in motion. The knowledge of the different forms of motion is the knowledge of bodies. The investigation of these different forms of motion is therefore the chief subject of natural science . . . Seated as you are there at the centre of the natural sciences you will be in the best position to judge if there is anything in it. (xx.646–50, 666; SC 281–2).

Marx's reply to this was friendly, brief and non-committal: 'Have just received your letter which has pleased me greatly.

But I do not want to hazard an opinion before I've had time to think the matter over and to consult the "authorities"² (xxxiii.82).

The 'authorities', so far as we know, did not seem to have been very impressed with Engels's insights, though Marx tried to break this to him gently. The chemist Carl Schorlemmer, for example, in marginal notes on Engels's letter, remarked that he agreed that the investigation of different forms of motion is the chief subject of natural science and that motion of a single body must be treated relatively ('Quite right!'). But when Engels wrote that dialectics, as *the* scientific world-view, could not itself advance from chemistry to 'organic science' until chemistry itself did so, and when he said with respect to biology, 'Organism – here I will not enter into any dialectics for the time being', Schorlemmer then commented, 'Me neither.' Marx's 'authority' found the science in Engels's letter more agreeable than the dialectics (xxxiii.80–1, 82, 84).

There was no more correspondence, so far as we know, between Marx and Engels concerning the *Dialectics of Nature* until Engels's letter of 21 September 1874, in which he commented that articles by Tyndall and Huxley in *Nature* had 'thrown me . . . back onto the dialectical theme', though on several occasions Marx referred to Engels's project and even made brief inquiries for him (xxxiii.119–20).

In the last exchange on Engels's research for the *Dialectics of Nature* Marx was very brief indeed. On 23 November 1882 Engels wrote:

Electricity has afforded me no small triumph. Perhaps you recall my discussion of the Descartes–Leibniz dispute . . . *Resistance* represents in electricity the same thing that *mass* does in mechanical motion. Hence this shows that in electrical as [in] mechanical motion – here speed, there strength of current – the quantitatively measurable form of appearance of that motion operates, in the case of a simple transition *without* change of form, as a simple factor of the first power; but in transition *with* change of form [it operates] as a *quadratic* factor. This is a general natural law of motion which I have formulated for the first time. (xxxv.118–19)

Marx's reply of 27 November was characteristically very much more specific, omitting any mention of natural laws: 'The confirmation of the role of the *quadratic* in the transition of energy with a change of form of the latter is very nice, and I congratulate you' (XXXV.120).

Engels elaborated the Marxism of the 1870s and 1880s in further works covering materialist philosophy and materialist accounts of the origin of man and his social and political institutions. In 1886 he seized an opportunity to clarify the presuppositions of the 'Marxist world outlook' and in so doing to finish the work started by Marx and himself in *The German Ideology*. Engels introduced his lengthy review of K. N. Starcke's *Ludwig Feuerbach* as 'a short, coherent account of our relation to the Hegelian philosophy', and 'a full acknowledgement of the influence that Feuerbach, more than any other post-Hegelian philosopher, had upon us during our period of storm and stress'. Dismissing the manuscript of *The German Ideology* as unusable, since it contained no criticism of Feuerbach's doctrine itself, and an incomplete exposition of 'the materialist interpretation of history', Engels did draw attention to Marx's hitherto unpublished eleven theses on Feuerbach, which he then added (in an edited form) as an appendix when his long review appeared as a book in 1888 (II.358, 359). In doing this Engels launched the first inquiry into the early Marx, tracing influences upon him, primarily philosophical, and searching in the early works for enlightenment concerning the origins and meaning of the later ones.

In revealing the 'true significance' of Hegelian philosophy – 'that it once for all dealt the death blow to the finality of all products of human thought and action' – Engels moved on in his *Ludwig Feuerbach and the End of Classical German Philosophy* to gloss once again Marx's 1859 preface to *A Contribution to A Critique of Political Economy* (II.362). Marx had written:

In broad outlines Asiatic, ancient, feudal, and modern bourgeois modes of production can be designated as progressive epochs in the economic formation of society. The bourgeois relations of production are the last antagonistic form of the social process of production – antagonistic not in the sense of individual antagonism, but of one arising from the social conditions of life of the individuals; at the same time the productive forces developing in the womb of bourgeois society create the material conditions for the solution of that antagonism. This social formation brings, therefore, the pre-history of human society to a close. (1.363–4)

In Engels's version 'all successive historical systems are only transitory stages in the endless course of development of human society from the lower to the higher'. 'Each stage,' he wrote, 'is necessary'. Engels's 'dialectical philosophy' was 'nothing more than the mere reflection of this process in the thinking brain'. This 'mode of outlook' was 'thoroughly in accord with the present state of natural science, which predicts a possible end even for the earth'. Though 'for the history of mankind', according to this dialectical view, 'there is not only an ascend-ing but also a descending branch', we were fortunately a 'con-siderable distance from the turning point'. Engels's method was to pursue 'attainable relative truths along the path of the posi-tive sciences, and the summation of their results by means of dialectical thinking' (11.362, 363, 365).

Having attempted to establish the relation of dialectical thinking to history and science, Engels tackled the 'great basic question of all philosophy': 'the relation of thinking and being'. In pursuit of that problem, he attempted to deal with the rela-tions of matter and consciousness, and scientific method and explanation:

We comprehended the concepts in our heads once more materialis-tically – as images of real things instead of regarding the real things as images of this or that stage of the absolute concept. Thus dialec-tics reduced itself to the science of the general laws of motion, both of the external world and of human thought – two sets of laws which are identical in substance, but differ in their expression in so far as the human mind can apply them consciously, while in nature and also up to now for the most part in human history, these laws assert themselves unconsciously, in the form of external necessity, in the midst of an endless series of seeming accidents. Thereby the dialectic of concepts itself became merely the conscious reflex of the dialec-tical motion of the real world and thus the dialectic of Hegel was placed upon its head; or rather, turned off its head, on which it was standing, and placed upon its feet. (11.387)

In the first version of this very famous and extraordinary metaphor Engels had written (in his 1859 review) that in Hegel's dialectic 'the real relation was inverted and stood on its head'. In 1872 Marx offered his own very spare comments on the method of *Capital* and its critical, corrective relation to Hegel's method. Marx observed that his 'dialectical method' was 'opposite' to Hegel's, because in Hegel's view 'the real world is only the external phenomenal form of "the Idea"', whereas his own view was the reverse: 'the ideal is nothing else than the material world reflected by the human mind, and translated into forms of thought'. While stopping short of Engels's theories of dialectical laws as the same for nature, history and thought, and of Engels's view that dialectical motion has its conscious reflex in the brain, Marx did comment that with Hegel the dialectic 'is standing on its head'. 'It must be turned right side up again', he wrote, 'if you would discover the rational kernel within the mystical shell' (c 1 29), rather as Engels had earlier claimed that Marx undertook the work 'of extracting from the Hegelian logic the kernel which comprises Hegel's real discoveries' (1.372, 373). Engels's apparatus of inversion, head-standing, kernels and shells defied even his own attempts to make sense of it, and apparently led Marx into a foggy realm of mixed metaphor.

In his own *Ludwig Feuerbach* Engels justified his general laws of motion by appealing to 'three great discoveries' in natural science: the discovery of the cell, leading to a 'single general law' of the development of all higher organisms and species; the transformation of energy as the manifestation and conservation of 'universal motion'; and Darwin's 'proof' that organic products, including man, were the result of evolution. Though the development of society differed in one respect, according to Engels, from that of nature (because in the latter there are only 'blind, unconscious agencies'), the conscious actors in society – who may be important in 'single epochs and events' – produced a 'state of affairs entirely analogous to that prevailing in the realm of unconscious nature'. In both nature and history, Engels wrote, 'accident holds sway' on the surface, but both were 'always governed by inner, hidden laws'. With that knowledge the course of recent history could be revealed in terms that were allegedly causal:

> But while in all earlier periods the investigation of these driving causes of history was almost impossible – on account of the complicated and concealed interconnections between them and their effects – our present period has so far simplified these interconnections that the riddle could be solved. Since the establishment of large-scale industry, that is, at least since the European peace of 1815, it has been no longer a secret to any man in England that the whole political struggle there turned on the claims to supremacy of two classes: the landed aristocracy and the bourgeoisie [middle class] . . . In modern history at least it is, therefore, proved that all political struggles are class struggles, and all class struggles for emancipation, despite their necessarily political form – for every class struggle is a political struggle – turn ultimately on the question of *economic* emancipation. (II.389, 390, 391, 393, 394)

In the 'Marxist conception of history', according to Engels, 'interconnections' were discovered 'in the facts'. Philosophy, 'expelled from nature and history', had left to itself only 'the realm of pure thought', which was 'the theory of the laws of the thought process itself, logic and dialectics' (II.400–1).

Engels's assertion of necessity in historical events was merely that; Marx had simply referred to successive epochs as progressive. How the brain could find its way from the realm of accident in thought to a reflection of dialectical development was similarly not explained by Engels; indeed the interrelationship of his categories of causation and accident, whether in the material world, in historical events, or in human cognition, was never explored. With respect to logic and philosophy Engels left us only his three dialectical 'laws' – quantity into quality, interpenetration of opposites, and development through contradiction (or negation of the negation) – together with his view that categories do not have fixed, unambiguous referents. Engels identified the latter view as the 'great basic thought' behind the 'materialist dialectic', writing that

> the world is not to be comprehended as a complex of ready-made *things*, but as a complex of *processes*, in which the things which are apparently stable, no less than their mental images in our heads, the

concepts, go through an uninterrupted change of coming into being and passing away, in which, in spite of all seeming accidentality and all temporary retrogression, a progressive development asserts itself in the end. (II.387)

However useful as principles of explanation and analysis, Engels's 'laws' and his overall dialectical view could not count as even the rudiments of a logical system.

Besides working out the 'materialist' presuppositions for knowledge of nature and history Engels also developed a 'materialist' account of the origin of man. Marx himself had made some observations on how exactly man differs from the animals when he discussed in *Capital* the concepts of labour and social production.

We are not now dealing with those primitive instinctive forms of labour that remind us of the mere animal. An immeasurable interval of time separates the state of things in which a man brings his labour-power to market for sale as a commodity, from the state in which human labour was still in its first instinctive stage. We presuppose labour in a form that stamps it as exclusively human. A spider conducts operations that resemble those of the weaver, and a bee puts to shame many an architect in the construction of her cells. But what distinguishes the worst architect from the best of bees is this, that the architect raises his structure in imagination before he erects it in reality. (C I 173–4)

Marx's discussion was more generally conceptual and abstract than Engels's quasi-historical speculations, such as the manuscript work, 'Labour in the Transition from Ape to Man'. This was written in 1876 but published just posthumously as an article in 1896. In it Engels undertook his own elaboration of labour as 'the prime basic condition for all human existence'. When apes 'walking on level ground began to disaccustom themselves to the aid of their hands and to adopt a more and more erect gait', they made *the decisive step in the transition from ape to man*'. Engels's view of evolution was Lamarckian, rather than strictly Darwinian, in that he believed that characteristics acquired by individuals could be inherited by later generations.

Thus the hand is not only the organ of labour, *it is also the*

product of labour. Only by labour, by adaptation to ever new operations, by inheritance of the thus acquired special development of
muscles, ligaments and, over longer periods of time, bones as well,
and by the ever renewed employment of this inherited finesse in
new, more and more complicated operations, has the human hand
attained the high degree of perfection that has enabled it to conjure
into being the paintings of a Raphael, the statues of a Thorwaldsen,
the music of a Paganini. (II.80, 81–2)

Darwin's supposed 'law of correlation of growth' was invoked speculatively by Engels to account for other instances of
variation and selection:

The gradually increasing perfection of the human hand, and the
commensurate adaptation of the feet for erect gait, have undoubtedly, by virtue of such correlation, reacted on other parts of
the organism. However, this action has as yet been much too little
investigated for us to be able to do more here than to state the fact
in general terms. (II.82)

According to Engels, labour began with the making of tools,
the most ancient of which were for hunting and fishing, and
this marked the transition 'from an exclusively vegetable diet to
the concomitant use of meat'. While paying 'all respect to the
vegetarians' Engels suggested that a meat diet was essential to
the rapid development of the brain. The development of social
production among humans was sketched by Engels to the point
at which we were 'gradually learning to get a clear view of . . .
our productive activity', so that after a complete revolution the
possibility will be afforded us of controlling and regulating its
effects (1.84, 85, 91).

Modern discoveries in anthropology, like those in biology,
chemistry, physics, history and philosophy reaffirmed, according to Engels, the truth of his 'materialist conception of history'
grounded in 'dialectics'. *The Origin of the Family, Private
Property and the State* was written and published in 1884 and
represented a lengthy attempt to demonstrate the compatability
of recent anthropological works with his 'materialist' account of
the origins of man and society, so that the latter would appear
confirmed by independent research. Engels was chiefly concerned with *Ancient Society*, published in 1877 by the Ameri-

movement, and particularly the influence of Bakunin. Engels's own definition of authority as 'the imposition of the will of another upon ours' presupposed, so he said, 'subordination', which was admittedly 'disagreeable to the subordinated party'. Even after a social revolution large-scale industry would still require a certain subordination, a certain authority. These were things 'imposed on us', he wrote. In keeping with his version of the dialectic he claimed that authority and autonomy were 'relative things' and that in the social organisation of the future, authority would be restricted 'to the limits within which the conditions of production render it inevitable'. Engels put his vision of this aspect of future society within his own characteristically broad dialectical perspective.

If man, by dint of his knowledge and inventive genius, has subdued the forces of nature, the latter avenge themselves upon him by subjecting him, in so far as he employs them, to a veritable despotism independent of all social organisation. (1.636, 637, 638)

Engels's dialectical perspective has been enormously influential, almost certainly because of its claims to encompass scientifically all the disciplines of physical and social studies within a single science. According to Engels this science predicts the 'inevitable downfall' of capitalism and justifies a political programme of emancipation for the workers of the world (II.135).

An assessment of Engels's work and its influence is now in order.

7 Engels and Marxism

The materialist interpretation of history is the main item in the intellectual legacy left us by Engels. These few thoughts, variously expressed by Engels himself, have had a revolutionary effect on social theory and political practice. An education in any of the arts or social sciences today with any claim at all to adequacy must include some consideration, however critical, of this doctrine. None of the attempts to show it to be vacuous, incoherent, tautological or illogical has succeeded, even when these attacks have been mounted by philosophers of the very highest reputation. The reason is that the materialist interpretation of history is so useful.

In practical politics a multitude of groups, including the Leninist, Trotskyist and Maoist strands of Marxist political activism, take the materialist interpretation of history as their first article of belief. Indeed, if there is a single criterion for determining who is a Marxist and who is not, the materialist interpretation of history would be the strongest contender. Mere acceptance of that conception would not, however, make anyone a Marxist in a very strong sense; anyway, to attach the label 'Marxist' to someone may not tell us very much, since there is no unitary interpretation of this famous view of history on which all Marxists agree. Rather the materialist interpretation of history represents a set of *shared disagreements*.

While in the political world the materialist interpretation of history functions as an article of belief (the primary point of justification for strategy, tactics and policy), the utility of this view appears more directly in works of history, sociology, political science, anthropology and philosophy. Both Engels and Marx himself credited *Marx* with a crucial insight into the nature and development of human society. Why then is 'the materialist interpretation of history' something left to us by Engels?

The first reason is that he invented the label itself. This

phrase became an object of exegesis independent of the complexities it was originally intended to summarise. 'Materialist', 'interpretation' and 'history' acquired a significance of their own independent of Marx's 'Theses on Feuerbach', the *Poverty of Philosophy* and, most importantly, his preface to *A Contribution to A Critique of Political Economy* of 1859. These terms did not fit Marx's view very well anyway. 'Production theory of social change' would be better than 'materialist interpretation of history', though Marx wisely refrained from calling his views anything. He only rarely referred to himself as a 'materialist', and then did not specify what this was intended to indicate except that he was not an 'idealist'. His 'Theses on Feuerbach' referred critically to previous materialisms and favourably to a 'new' materialism, though Marx did not connect this label with anything more specific than 'human society, or social humanity'. While Marx had views on the historical development of capitalist society, his was not a task of 'interpretation'. 'The philosophers have only *interpreted* the world', he wrote disparagingly in the eleventh thesis on Feuerbach. Nor was he really concerned with 'history' as a historian would be in producing an 'interpretation'. Marx aimed at a *revolutionary practice* – 'self-change' in human society (5.4, 5).

The second reason for saying that the materialist interpretation of history was a legacy from Engels is that he was its original and to date its most effective exponent. He was not merely a labeller but also a *glosser*. This proved to be the most important explicatory technique used by Engels in all his writings, because it is his glosses on Marx that have been most influential, rather than any of his historical researches or contemporary observations.

In his glosses on Marx, Engels's intentions, so far as I can tell, were wholly honest and honourable. He quoted with reasonable accuracy and gave credit where it was due. While his own political and intellectual reputation was enhanced by his relationship with Marx and his interpretations of the master's works, he kept his claims and ambitions within the bounds of discipleship.

Marx's 1859 preface to *A Contribution to A Critique of*

Political Economy contained the few paragraphs of his 'guiding thread'. Understandably this text became the prime object for Engels's glosses, his running commentary on Marx's thoughts, and his expansion of his views. The first gloss was crucial in setting the method and content for the later ones.

After quoting Marx extensively in the 1859 review of *A Contribution to A Critique of Political Economy*, Engels moved on to state what he took to be the essence of Marx's views and to expand them, rather as if he were reworking the jointly written *German Ideology* of 1845–6 on his own. In that way, perhaps, the notion of joint authority over the development of these views grew in his mind. The thought that what he said in his gloss might, however slightly, conflict with Marx's insights never seems to have occurred to him. As we have seen, Marx and Engels were joint authors of only three important works, all written before 1850. After that, Engels's works were published under his own name, and Marx took no responsibility for them in practice. Engels, however, did not see the matter in quite that light, though the assumption of joint authority was not explicitly publicised until after Marx's death in 1883. By then Engels was inescapably tied to the implications of his 1859 gloss on the 'guiding thread' enunciated by Marx in his 1859 preface.

Engels's gloss in his 1859 review contained a move that was to prove crucial for the history of Marxism. Fired by the certitude of Marx's formulation in *A Contribution to A Critique of Political Economy* of the laws of capitalist society, Engels projected those claims to rigour and precision on to Marx's much less strict account in his preface of the general nature of society and its general pattern of development. In those passages Marx spoke in terms of correspondence, conditioning and determination (i.e. definition, limitation) – not in terms of every 'action' originating from 'material impulses'. Marx wrote:

In the social production of their life, men enter into definite relations that are indispensable and independent of their will, relations of production which correspond to a definite stage of development of their material productive forces. The sum total of these relations of production constitutes the economic structure of society, the real foundation, on which rises a legal and political superstructure and to which correspond definite forms of social consciousness. The mode of production of material life conditions the social, political and intellectual life-process in general. It is not the consciousness of men that determines their being, but, on the contrary, their social being that determines their consciousness. (1.362–3)

The passage just quoted was at a much higher level of generality than the laws of capitalism formulated by Marx in his published and unpublished works from 1859 onwards. Those laws were precisely and confidently stated: the law of value, the law of the tendency of the rate of profit to fall, 'the economic law of motion of modern society', as he put it in the preface to the first edition of the first volume of *Capital* (c 1 20).

Contrary to this interpretative move, Engels employed the general observations of the 1859 preface in much the way Marx did. His own historical work reflected this. Ideas, doctrines, movements and parties were shown in his accounts to have eminently practical relationships with the control and division of resources, with economic life in all its aspects. This has been the most influential idea of modern times in the study of man and society and in the practical alteration of political and economic life around the world.

It was a short step from this projection of rigour on to Marx's 'guiding thread' to the point at which Engels labelled it a 'law' and made claims about its universality and certitude that were not made by Marx. Engels called his law the 'great law of motion in history', analogous in scope and precision with 'the law of the transformation of energy' (1.246). This claim was patently untrue. Engels took his 'materialist' account of the formation of classes and the development of society to be concerned with *ultimate* economic causes, as Marx did not, and he took those economic causes to be connected (somehow) with the materialism of the physical sciences. Even Marx's most rigorous 'laws' of capitalism were never linked with matter in motion. Neither the doctrine of ultimate causation nor that of the connection of economic phenomena with matter as conceived by physical scientists was ever explored in Engels's works, and was therefore certainly not explained and justified. Engels's own three laws of dialectics did not help in this task, since they have never been accepted by physical scientists as intrinsic to science. Nor were they, in any case, testable propositions, since it was not clear in Engels's account what was and what was not an instance of their operation. Engels's formulae simply did not have the general character, but precise referents, of, for example, Newton's laws of motion and Boyle's law concerning the behaviour of gases.

Engels might have recommended Marx's 'guiding thread' as a *hypothesis* for investigating historical and contemporary conflicts in society. A hypothesis may of course not pay off in every investigation of every conflict. Marx did not in the 1859 preface assert that all individual actions and social conflicts would be effects in some traceable sense of the mode of production of material life. Engels departed from Marx in claiming that he had found a historical law in accord in some ultimate causal sense with all events. Moreover, by interpreting 'material life' to imply the materialism of the physical sciences, he glossed Marx's views on men and their material productive activities out of all recognition.

Attempts to defend his 'materialist interpretation of history' from critical opponents and from naïve practitioners became, late in life, an increasing preoccupation for Engels. In 1890 he wrote these now famous lines to a correspondent:

According to the materialist conception of history, the *ultimately* determining element in history is the production and reproduction of real life. More than this neither Marx nor I have ever asserted. Hence if somebody twists this into saying that the economic element is the *only* determining one, he transforms that proposition into a meaningless, abstract, senseless phrase. (SC 417)

Engels then detailed what other factors were operative in historical events:

The economic situation is the basis, but the various elements of the superstructure – political forms of the class struggle and its results, to wit: constitutions established by the victorious class after a successful battle, etc., juridical forms, and even the reflexes of all these actual struggles in the brains of the participants, political, juristic, philosophical theories, religious views and their further development into systems of dogmas – also exercise their influence upon the course of the historical struggles and in many cases preponderate in determining their form. (SC 417)

Engels's defence of his 'materialist interpretation of history' was analytically indeterminate and ultimately dogmatic, because interaction between base and superstructure was never distinguished from ultimate causation by the base, which was in turn never successfully squared with the general account of economic and intellectual life given in *The German Ideology*. It would take a great many distinctions and a good deal of argument and example to elucidate and justify his confident but vague assertion: 'There is an interaction of all these elements in which, amid all the endless host of accidents . . . the economic movement finally asserts itself as necessary' (SC 417). The arguments behind Marx's thesis that social, political and intellectual life were conditioned by the mode of production derived from his view that living individuals in material surroundings must produce their means of subsistence and that this had a defining and delimiting effect on their culture. Engels equated Marx's anti-idealism with a composite but incomplete materialism: a view that matter in motion accounts for everything, *and* a view that human beings must grapple with the material conditions of production, both those they find and those they make. There is no doubt that Marx held the latter view, but Engels's preoccupation with the former, which was, for all his protestations, a materialism of the traditional sort, has led to further glossing of Marx. Hence it has become an interpretative commonplace that the Marxian base and superstructure were mutually exclusive categories in the specific sense that the economic structure or base was somehow *material* and the superstructure wholly *immaterial*, consisting of ideas. Since under 'relations of production' Marx had in mind human economic activities which obviously required both ideas *and* material things, this interpretation of the base–superstructure distinction was somewhat strained, to say the least, and the apparent contradiction of including immaterial factors in the base arose solely from the habit of commentators of assuming that Marx's new materialism must be of the sort described

by Engels, namely matter in motion. Superstructural phenomena (Marx mentions law, politics, religion) were also obviously mixtures of *material* factors and consciousness, as
indeed was human life itself, on his view. Marx was unin-
terested in the matter–consciousness dichotomy; Engels, by
contrast, was only too willing to presuppose this traditional
philosophical move in considering human experience, and to
assert with confidence that the two were related in some ultimate and dialectical sense which he never satisfactorily
examined and specified.

Though the materialist interpretation of history cannot be
defended successfully as a causal law in history, still less as a
law derived from the materialism of the physical sciences, it has
proved its utility beyond doubt as a *hypothesis* in accounting
for social change, a guide to research that leads, more often than
not, to important results in the study of human society. A
hypothesis about social life need not be true or even apposite
with respect to *every* social event. Rather it provides a starting-
point for investigations. However untrue or inapposite it may
prove to be for a given event, its potential utility in explaining
other events is unaffected. If it never worked we would reject it;
many times, however, it does work, sometimes brilliantly.

In my assessment of the chief item in Engels's intellectual
legacy – the materialist interpretation of history – I have tried to
sharpen the debate among Marxists and non-Marxists alike, the
debate about what it says and means, and about its truth and
utility. My method has been to draw attention to Engels's role
as glosser of Marx's texts, and to the content of his glosses,
noting the points at which I think the glosses deviate significantly from the original material, and the new problems that
this creates. Those who accept the substance of Engels's glosses
have run into considerable difficulty explicating and justifying
his concepts of causality in the physical world and in social life.
This has led to debates about free will and determinism, and
these in turn to difficulties in justifying political initiative. Some
commentators have suggested that the influence of Engels's
philosophical views on the Second International, the world-
wide organisation of socialists that functioned from 1889 until

the First World War, was disastrous. According to this account, Engels's causal determinism encouraged certain socialist leaders to act as if proletarian revolution would simply arrive as history took its course, so that their adherence to revolutionary principles could remain largely formal. While Engels can hardly be held accountable for the decisions of others, the lack of clarity in his accounts of ultimate causation in the materialist interpretation of history contrasted with his own consistency in revolutionary politics, and made it easy for some socialists to entertain ambiguous notions of historical inevitability and the dialectic of history.

I have claimed Engels as the first Marxist historian and anthropologist, and here his influence has led to results accepted as progressive within those disciplines. His own writing, linking political events to social classes and the economic structure of society, stood apart from his methodological prescriptions and analysis. His works of history and anthropology contained insights and hypotheses that have stimulated further research on the subjects that interested him, and on others besides.

I have also suggested that Engels was the first to turn to the early works of Marx, including his notebooks, for enlightenment on the substance, and particularly the premises, of his mature works. This was an instance of Engels's genuine intellectual interest in glossing Marx as fully and informatively as possible. But at the same time this development reflected Engels's inability to deal in comparable detail with the more overtly economic works of the later Marx, and with the detailed exposition they presented. In a sense, Engels left economics to Marx. Whether Marx knowingly left anything to Engels to do – and it was often claimed that, for example, natural science, philosophy and military affairs were his domain by mutual agreement – has not been revealed in the documentation we have.

While Marx's early works are a fascinating object of study, and while there is enlightenment in them about Marx's premises in the mature works, Engels perhaps unwittingly set a trend among students of Marx that has led to the neglect of *A Contribution to A Critique of Political Economy* and *Capital* in favour of a rather backward-looking debate: an assessment of the battles of the 1840s over idealism, materialism, Hegel and Feuerbach. From his reconsideration of these early debates Engels produced another gloss on Marx's work: the concept of false consciousness, as described in his letter of 14 July 1893 to Franz Mehring, who was later Marx's biographer. 'Ideology is a process accomplished by the so-called thinker consciously, it is true, but with a false consciousness. The real motive forces impelling him remain unknown to him; otherwise it simply would not be an ideological process. Hence he imagines false or seeming motive forces' (SC 459). The subtleties of Marx's early analyses of idealist philosophy, religion, law and capitalist apologetics were almost wholly obscured by Engels's blanket notion of falsity, and his unexamined concept of consciousness. Engels did an immense service, however, in presenting the second and third volumes of *Capital* for publication with very little gloss on the substance of Marx's masterpiece.

Some of the critical and analytical categories that have been applied by commentators to Marx actually fit Engels rather better. In reading Engels's works, one has more of the feeling, and considerably more persuasive evidence, that one is looking at someone who was the object of successive influences, than with Marx. It was Engels who wrote full-blown Hegelian and then Young Hegelian prose in the early 1840s. It was Engels who adopted the communist perspective swiftly, and as a whole. It was Engels who took the views of a mentor to be established laws, proved beyond doubt. And it was Engels who adopted a positivist view of natural science projecting it on to Marx and Darwin alike, when in fact it suited neither.

Marx always had his own critical perspective, and he kept his admiration for the various authorities whose views he considered under much firmer control than Engels. This has been somewhat obscured by the notion of 'influence'. From 1840 onwards, one would never really mistake Marx's work for that of a Hegelian, Feuerbachian, Young Hegelian, Ricardian or positivist, however much he might actually have been in agreement with any of these authors or schools of thought. The same could not always be said of Engels. While it has sometimes been claimed that Marx moved from philosophy to economics, and then to positivism in social science, that transition actually describes Engels's career rather more accurately, since Marx was firmly focused on 'so-called material interests' and political economy from 1842 onwards (1.361–2). Still, Engels was rather better than Marx at eyewitness accounts, short statements of principle, and easily readable polemics and popularisations.

Engels's thought as a whole reflected certain fashions in nineteenth-century philosophy, among them system-building in the manner of Hegel and Dühring, the materialism and determinism of the physical sciences, evolutionism derived from Darwin, atheism arising from historical criticism of theology, and positivism in the view that theory arose from fact. Unlike Marx, who used some of these materials in a strikingly original and critical way, Engels was an autodidact and lacked the sophistication of a trained sceptic who could put awkward questions to himself and then strive, painfully, to answer them. Engels's philosophy was not merely scattered through various polemics, like Marx's, but was in itself a meagre body of work, with many unexamined assumptions, undefined terms and unspecified relationships.

The efforts of Marx and Engels to establish themselves as a political influence helped to ensure that their works would be read in future, irrespective of their utility as contributions to the social sciences. With respect to philosophy, history, sociology and the other arts and sciences, it was Marx who made the more original and Engels the more influential contributions, chiefly in his late works. Engels's sweeping claims concerning the sciences and their relation, properly understood, to political activity, were crucial for this. So was his strong version of Marx's 'guiding thread' – a projection of Marx's certitude concerning the laws of capitalism on to his more general formulations concerning the overall nature and development of society up to the age of industrialisation and beyond. To that projection of certitude Engels added his view that economic causation was in some unspecified sense analogous to causation in the physical sciences. The young Engels was actually closer to Marx's premises, as Marx himself acknowledged, since the sociological works, culminating in *The Condition of the Working Class in England*, demonstrated the defining and delimiting character of the new industrial society upon law, politics and cultural life. Had Marx never existed, that work would no doubt still be read. Engels's historical writings from his middle period, about 1850–70, deserve a much wider audience than they have today. His accounts of the peasant war in Germany and of the 1848–9 revolutions, and the articles on warfare and military developments in Europe and America, have generally been read by the converted as instances of Marxist historical analysis, which they are. They should, however, be more widely studied and evaluated.

I have said little about Engels's personal life, since I think that most of it was irrelevant to a brief consideration of his ideas. It is true that he successively kept the working-class Burns sisters, Mary and Lizzie, as his mistresses in accommodation in Manchester separate from his own lodgings as a middle-class bachelor. And it is true that he rode to hounds, had a taste for champagne and claret, and maintained himself in fashionable neighbourhoods and spas. Some commentators have implied that there was an incongruity between his life-style and his political position as a revolutionary communist. But had Engels been respectably married, taken little exercise, lived in poverty or with very modest means, eschewed paid employment and resided in lower-middle-class surroundings, I doubt whether he would have escaped this particular censure any more than Marx (whose circumstances I have just described). Had Engels and Marx lived impeccably proletarian lives they would probably have had no time for their intellectual labours, and in any case subsequent critics might have attacked them for belieing their own middle-class origins and becoming phonies. The life-style of any radical critic of contemporary social arrangements is bound to look incongruous.

The story is current that on his deathbed in 1895 Engels revealed that Marx was the father of Frederick Demuth, the son of Marx's housekeeper. The sole evidence for this deathbed revelation is what seems to be a copy (whose provenance is unknown) of a letter from Engels's former housekeeper Louise

Freyberger written in 1898. While some commentators see no reason to doubt the authenticity and accuracy of this document and the truth of what Engels is supposed to have said, others have suggested that there are internal inconsistencies in the supposed letter that throw doubt on its being a genuine copy. However, even if we did accept the copy as genuine and the account of Engels's remarks as accurate, there are still grounds for scepticism, since what Engels was claiming is otherwise uncorroborated. Research into the life of Frederick Demuth and of his relations has yielded nothing concerning the identity of his father; letters in the Marx–Engels collection from the period of Frederick Demuth's birth and subsequent life do not establish anything definite about the situation, and nothing else about him is known that would link him to Marx, though unsubstantiated claims have been made. I mention this matter to draw the reader's attention to a subject which, so far as I know, has no bearing on Engels's work, but is worth some consideration as an indication of the state of scholarship on Engels.

Engels sometimes made statements in articles and correspondence employing racial categories. While it might be possible to show that he had what could be characterised today as racist views, it is wholly inaccurate to claim that his philosophical work, or indeed his intellectual legacy and influence, were in any significant sense racist or even favourable to racism. If he had views that we would term racist today, he held them independently of his Marxist outlook, where racial categories did not figure.

The Marx–Engels intellectual relationship emerges, in my account, as one of mentor and glosser. Except for the brief flurry of joint projects in the 1840s, the two seem to have worked independently on their major theoretical pronouncements. The requests for assistance and announcements of discoveries in the correspondence that survives do not support the claims, commonly made, that Engels and Marx were completely at one on all issues and that they functioned as joint authors, each taking the other's work as his own, each seeing the other as partner in a collective venture. Rather the picture that emerges is of work undertaken independently and separately pursued, with minor exceptions. Some of the requests for assistance and approval produced no replies; some drew only brief, non-committal responses. The two could not have taken the stance of joint authorship and joint responsibility in their private meetings and have written the letters that survive. These letters do not support the view that Marx and Engels functioned as a perfect intellectual partnership. But in their correspondence, the subjects of historical research, political news, family gossip and party affairs were a different story, and on those topics we have a record of lively interchanges between separate but allied personalities.

Engels himself initiated the view that he and Marx were in agreement on all fundamentals – fundamentals that then emerged in Engels's glosses on Marx – and that the joint authority of 'Marx and I' could be invoked in setting out the 'materialist interpretation of history' and other doctrines. After Marx's death Engels recommended his own works, such as *Anti-Dühring* and *Ludwig Feuerbach*, to be read alongside Marx's, though he said rather more strongly to one correspondent in 1890 that while *Capital* merely alluded to 'historical materialism', 'I have given the most detailed account' (SC 418).

Commentators, adherents and critics were not slow to seize the enormous advantages offered by this view of the Marx–Engels relationship. The style and content of Marx's works were more difficult, particularly in the critical works on political economy, than Engels's more readable efforts; indeed Engels's subjects – philosophy and history – were less remote than political economy. There were some aspects of Engels's work that were easier to demolish than Marx's more intricate arguments, so hostile critics have clung to the view that Marx and Engels may be read interchangeably. Political and academic life in the official institutions of the Soviet Union, by contrast, involves a positive commitment to dialectical and historical materialism that derives from Engels's works but requires the posthumous imprimatur of Marx, the senior partner. The Marx–Engels relationship is therefore sacrosanct.

Some Western commentators, while suspecting or acknowledging important differences between Marx and Engels, have chosen to ignore the matter, usually dealing with Marx alone. Others have accepted the view that Marx and Engels spoke for each other, and then defended Engels's glosses on Marx independently of Marx's texts, or in some cases attempted to demonstrate that Marx's texts agreed with Engels's. No one, so far as I know, has tried to demonstrate that Engels's causal laws are of the same high order of generality as Marx's formulations in the 1859 preface.

Perhaps the most recent influential view of the textual differences between Marx and Engels is that Marx drifted towards the positivism and determinism espoused in Engels's glosses without saying so explicitly. If this were true, then the high status accorded to Engels's works by many Marxists would have had the tacit approval of the master. This view is not, however, very well supported by what Marx actually said during his career. Laws of dialectics did not appear in his preface to *A Contribution to A Critique of Political Economy* of 1859, his popular work *Wages, Price and Profit*, his masterpiece *Capital* and associated manuscripts, nor in his last work of theoretical interest, his *Notes* on Adolph Wagner (an academic political economist).

The evidence usually cited for the view that Marx endorsed the 'materialist' theories of history and dialectics espoused by Engels is the *claim* that Marx approved of Engels's *Anti-Dühring* and agreed in principle with the manuscript *Dialectics of Nature*. But, as we have seen, it was only in the 1885 preface to the second edition of *Anti-Dühring* (written *after* Marx's death) that Engels publicised Marx's help in collecting material for the chapter on political economy. And it was only then that Engels claimed that he had 'read the whole manuscript' to Marx 'before it was printed'. We have no other evidence to support this story. Moreover, in the 1885 preface Engels also wrote that his 'exposition' of the 'world outlook fought for by Marx and myself' should not appear without Marx's 'knowledge'. This, Engels said, was 'understood' between them. He thus gave the reader the impression that Marx approved his work as an expression of 'their' outlook, while avoiding the statement that Marx agreed explicitly to any such thing (AD 13–14). There

were no recorded responses or revisions by Marx to the substance of Engels's work in *Anti-Dühring*. In fact Engels seems
to have made no move to put Marx's name on the book or to
gain and publicise an imprimatur.

However, if Marx were at odds with Engels over the substance of *Anti-Dühring*, why did he not dissociate himself from
it? Or had he never read it (or listened to it) in the first place?

Anti-Dühring was published so many times in 1877–8, even
before the widely circulated abridgement in *Socialism: Utopian
and Scientific*, that Marx could hardly have missed it. Engels
actually sent him an inscribed copy of the book. Even if Engels's
story about reading the manuscript to Marx were untrue, or if
Marx were not listening, it would be perverse to imagine that he
ignored the content of the work altogether. Perhaps he felt it
easier, in view of their long friendship, their role as leading
socialists, and the usefulness of Engels's financial resources, to
keep quiet and not to interfere in Engels's work, even if it con-
flicted with his own. After all, *Anti-Dühring* went out under
Engels's name alone.

Interestingly, Engels did not claim to have shown Marx the
Dialectics of Nature, which he interrupted in order to write
Anti-Dühring. In that work his views on the nature of dialectics were formulated explicitly, which was not quite the case in
the first edition of *Anti-Dühring*. Engels, it seems, had been
canny enough to avoid provoking disagreements with Marx
while the latter was still alive. And Marx seems to have been
similarly canny in not pressing Engels for details of his work.

In the event it was possible for Marx to take the view that the
first edition of *Anti-Dühring* would do more good than harm
within the socialist movement, since he detested Dühring's
views, and Engels savaged them relentlessly. Marx also recommended the book to others, referring very simply to Engels's
'positive developments' and to the political importance of *Anti-
Dühring* for 'a correct assessment of German socialism'. He did
not thereby commit himself to every philosophical and methodological implication of the text or to the view that it could
be read instead of *Capital*, a notion that Engels encouraged in
private (xxxiv.263–4, 346; xxxv.396). Least of all was Marx

committed to Engels's later glosses on *Anti-Dühring* or to what Engels subsequently claimed about the relationship between their independent works.

In their consistent espousal of a strategy of proletarian revolution Engels and Marx were rather more at one, though neither denied that reformism might have its place and bring successes. They simply saw too much room for failure in policies of moderation and compromise, and hence saw no virtues in reformism as such. Engels, like Marx, left little in the way of substantive political writing on party organisation, decision-making and leadership, by contrast with some of his successors in European Marxism, among them Lenin, Trotsky and Rosa Luxemburg. While it is possible that Engels's views on the ultimate nature of reality had a deleterious influence on revolutionary socialism, the case has not been proved and probably could never be established one way or the other, since the fate of socialism in the early twentieth century could hardly be set down to something so purely intellectual.

Engels left us a science that is obscurely comprehensive, unexamined in its determinism and old-fashioned in its materialism. Marx left us something rather different and rather more complex, though there is little agreement as yet on exactly what his critique of political economy means for contemporary social science and politics. In so far as Engels's interest in the premises of Marx's views has stimulated a rereading of his and their early works, the influence of his writings has been positive, though the need to separate Engels's glosses from Marx's text remains paramount. Marx's own life-work, however, was built on just those premises, and it is the material in *Capital*, edited but not substantially glossed by Engels, that stands as a challenge today.

In my account of Engels's thought I have tried to show that disagreements about the content of his work and its relationship to Marx's are not merely disputes about texts and intellectual biography but are about the substantially different approaches to social science and perhaps to politics itself that we find in their respective writings and careers. I have considered the substance of Engels's views and shown how in some cases they arose from his glosses on Marx. I have considered also the relationship of Engels's glosses to Marx's work itself and to what Marx actually had to say about Engels's efforts. And I have discussed further glossing of Engels's views and the effect of this on later interpretations of Marx's work, on Marxist politics and on our intellectual life, particularly in the social sciences. Social science incorporates what knowledge of society we have, and politics is our means of changing it. The theoretical and practical battles about Engels – his views, his works, his relationship with Marx – are far from over.

Further Reading

Works by Engels

The *Collected Works* of Karl Marx and Frederick Engels will present Engels's works and letters in English translation (or in the original English) in approximately 50 volumes when the series is completed. The first volume appeared in 1975, and the publishers are Progress of Moscow, Lawrence & Wishart of London, and International of New York, referred to below as the Progress consortium. All the major works of Engels (and the joint works with Marx) mentioned in the text are available in Progress editions, and *The Condition of the Working Class in England* is also translated and edited by W. O. Henderson and W. H. Chaloner (2nd ed., Blackwell, Oxford, 1971; Stanford University Press, 1968).

The *Selected Works* of Karl Marx and Frederick Engels in one volume was first published in 1968 and has been reprinted by the Progress consortium; of Engels's major works it includes *Socialism: Utopian and Scientific* and *The Origin of the Family, Private Property and the State*, with *Ludwig Feuerbach and the End of Classical German Philosophy*. The *Selected Works* in two volumes from the same publishers includes more of Engels's shorter writings, such as the 1859 review 'Karl Marx: A Contribution to the Critique of Political Economy' and 'On Authority'. *Engels: Selected Writings*, edited by W. O. Henderson (Penguin, Harmondsworth, and Baltimore, Md, 1967) contains selections from *The Condition of the Working Class in England* and the full text of the 'Outlines of a Critique of Political Economy', as well as other economic, historical, philosophical and military writings. *Engels as Military Critic*, edited by W. O. Henderson and W. H. Chaloner (Manchester University Press, 1959; Greenwood Press, Westport, Conn., 1976) presents a selection of lesser-known articles of the 1860s. *German Revolutions*, edited by Leonard Krieger (University of Chicago Press, 1968), includes

The Peasant War in Germany and *Germany: Revolution and Counter-Revolution.*

Works about Engels

I am indebted to the factual material collected and very well documented in W. O. Henderson's *The Life of Friedrich Engels* in two volumes (Frank Cass, London and Portland, Ore., 1976). Gustav Mayer's two-volume biography in German is published as *Friedrich Engels* in an abridged English translation by Gilbert and Helen Highet, edited by R. H. S. Crossman (Chapman & Hall, London, 1936; H. Fertig, New York, 1969). David McLellan's Modern Masters *Engels* (Fontana/Collins, Glasgow, 1977; Penguin, Baltimore, Md, 1978) presents a brief account of Engels's life and works.

Engels's major works are discussed in Fritz Nova's *Friedrich Engels: His Contributions to Political Theory* (Vision Press, London, 1968; Philosophical Library, New York, 1967). *Engels, Manchester and the Working Class* by Steven Marcus (Random House, New York, 1974) presents an analysis of the work from a literary point of view. The Marx–Engels relationship is considered in Norman Levine, *The Tragic Deception: Marx contra Engels* (Clio Books, Oxford and Santa Barbara, Calif., 1975). I am working on a study of the Marx–Engels intellectual relationship to be published by Harvester, Brighton, in 1982.

The relationship of Engels to Marxism is discussed in George Lichtheim's *Marxism: An Historical and Critical Study* (2nd ed., Routledge & Kegan Paul, 1968; Praeger, 1965), and in Richard N. Hunt, *The Political Ideas of Marx and Engels*, vol. 1, *Marxism and Totalitarian Democracy* (Macmillan, London, 1975; University of Pittsburgh Press, 1974). This topic is covered in three very recent studies: Leszek Kolakowski, *Main Currents of Marxism*, translated by P. S. Falla, vol. 1 (Oxford University Press, Oxford and New York, 1978); David McLellan, *Marxism after Marx* (Macmillan, London, 1979; Harper & Row, New York, 1980); and Alvin W. Gouldner, *The Two Marxisms* (Macmillan, London; Seabury Press, New York, 1980).

Four articles of interest in which Engels's work is discussed are: Terrell Carver, 'Marx, Engels, and Dialectics', *Political Studies*, vol. XXVIII, No 3 (September 1980), pp. 353–63; Gareth Stedman Jones, 'Engels and the End of Classical German Philosophy', *New Left Review*, No 79 (May–June 1973), pp. 17–36; the same author's 'Engels and the Genesis of Marxism', *New Left Review*, No 106 (November–December 1977), pp. 79–104; and Paul Thomas, 'Marx and Science', *Political Studies*, vol. XXIV, No 1 (March 1976), pp. 1–23. The last-named article has been particularly helpful to me in working out my views on Engels.

There is now an excellent Marx–Engels bibliography in English: Cecil L. Eubanks, *Karl Marx and Friedrich Engels: An Analytical Bibliography* (Garland Press, London and New York, 1977).

Index